T0309167

ROBUST METHODS
FOR DATA REDUCTION

ROBUST METHODS
FOR DATA REDUCTION

ALESSIO FARCOMENI
LUCA GRECO

CRC Press
Taylor & Francis Group
Boca Raton   London   New York

CRC Press is an imprint of the
Taylor & Francis Group, an **informa** business

A CHAPMAN & HALL BOOK

CRC Press
Taylor & Francis Group
6000 Broken Sound Parkway NW, Suite 300
Boca Raton, FL 33487-2742

© 2015 by Taylor & Francis Group, LLC
CRC Press is an imprint of Taylor & Francis Group, an Informa business

No claim to original U.S. Government works

Printed on acid-free paper
Version Date: 20150312

International Standard Book Number-13: 978-1-4665-9062-5 (Hardback)

Visit the Taylor & Francis Web site at
http://www.taylorandfrancis.com

and the CRC Press Web site at
http://www.crcpress.com

To Sara

To Simona

Contents

Preface

Robust statistics is concerned with statistical procedures leading to inference that is stable with respect to departures of the data from model assumptions. Data inadequacies often occur in the form of contamination: anomalous values, or *outliers*, are in disagreement with the data generating mechanism. As an example, consider point estimation procedures. Robust point estimation methods target the same quantities (location and covariance, for instance) as the classical ones, but ignoring or at least *downweighting* suspicious observations.

A great deal of work has been done on characterizing and defining robustness. Nowadays robust techniques have been developed in practically any field in statistical analysis. The milestones are books by Huber (1981), Hampel *et al.* (1986), Maronna *et al.* (2006), Huber and Ronchetti (2009).

Despite the considerable achievements from the theoretical point of view, in our experience at present, robust methods are not often used in practice. The aim of this book is to fill the existing gap between theoretical robust techniques and the analysis of real data sets in the area of data reduction. We would like to present the robust techniques with real examples and a clear practical motivation, so to encourage the dissemination and use of robust methods. In a sense, our book complements that of Heritier *et al.* (2009), which gives a very well-written account of robust methods for asymmetric modeling and statistical inference in general. A similar attempt to review and recommend robust inferential methods in medical research has been done by Farcomeni and Ventura (2012). We point the reader also to the excellent review of Hubert *et al.* (2008), and to the recent book of Varmuza and Filzmoser (2008).

Data reduction techniques are of primary importance in many fields, including economics (Brigo and Mercurio (2001), Wolfson *et al.* (2004)); genetics (Eisen *et al.* (1998), Reich *et al.* (2008)); psychology (Russell (2002), Clatworthy *et al.* (2005)); engineering (Apley and Shi (2001), Lu *et al.* (2003), Farcomeni *et al.* (2008), Prassas *et al.* (2007)); social sciences (Collins *et al.* (2004)); environmental research (Aires *et al.* (2000), Khatun (2009), De Sanctis *et al.* (2013)). An excellent introductory book on data reduction and multivariate analysis is Chatfield and Collins (1980).

Checking sensitivity with respect to model assumptions and evaluating the possible presence and influence of outliers is fundamental in all applications. In a surprising majority of cases, some or even substantial sensitivity can be found to modeling assumptions. Moreover, very few observations (even a single one) may have a strong and undesirable influence on classical inference. When the conclusions are too dependent on the validity of some of the assumptions, but data are not completely in agreement with them, robust methods must be used.

Research in robust statistics has provided alternatives to classical approaches (like principal component analysis, discriminant analysis, k-means) which have substantially the same interpretation and are resistant to the occurrence of data inadequacies, hence leading to valid inference even if model assumptions do not hold for all the data at hand. Robust analyses can be employed without even having to worry about sensitivity and automatically react to contamination. Hence, practitioners can use robust methods without even having to worry about contamination in the data. On the other hand, high breakdown point (that is, very robust) estimators pay a price in efficiency when there is no contamination in the sample. The user must therefore be aware of the relative efficiency of the tools used. In our opinion, robust analyses can and should be routinely used in practice, at least to confirm the results of classical analyses and evaluate their sensitivity to assumptions. In case some disagreement is found, robust analyses are always more reliable and protect us against being mislead by the data. Most robust procedures are now implemented into R and other software, which makes them readily available.

Another feature of robust statistics is that most procedures can be used to identify anomalous values to some extent. This is often a very important insight into the data generating process. When outliers are not arising simply as gross errors in data entry, they are *hypothesis generating*. In some cases, outlying measurements correspond to subjects belonging to a rare subpopulation in the data, which could be of interest for further investigations and future studies. In summary, robust methods give us reliable inference for the population of interest *and* an indication of a further population which had not been considered at the moment of planning the study.

Despite the fact that they could be extremely useful in applications, to the best of our knowledge robust data reduction techniques are the exception, rather than the rule, in almost all fields. Our aim is to give a non-technical review of robust data reduction techniques, in order to help dissemination and use of these important and useful methods. The main fields we deal with include principal components analysis e.g., Rousseeuw and Leroy (1987), Croux and Haesbroeck (2000), Salibian-Barrera *et al.* (2006), Maronna (2005), Hubert *et al.* (2005), Croux and Ruiz-Gazen (2005), Croux *et al.* (2007); sparse principal component analysis e.g., Croux *et al.* (2013); canonical correlation analysis e.g., Branco *et al.* (2005), Taskinen *et al.* (2006), Shevlyakov and Smirnov (2011); factor analysis e.g., Pison *et al.* (2003); clustering e.g., Kaufman and Rousseeuw (1990), Cuesta-Albertos *et al.* (1997), Gallegos and Ritter (2005), García-Escudero *et al.* (2008, 2010); double clustering e.g., Farcomeni (2009b); and discriminant analysis e.g., Hubert and Van Driessen (2004).

The book is divided in two parts, which follow two introductory chapters. The first introductory chapter gives a detailed motivation and an overview of the book. Some of the data sets which will be used to develop examples throughout the book will also be introduced and described in the first chapter. The second introductory chapter is a bit more technical and gives an overview of location and covariance estimators, and some outlier identification techniques.

We then proceed with the first part of the book, where we focus on *dimension reduction*. These techniques aim at synthesizing available information by reducing the dimensionality of the data. We begin with a chapter devoted to principal component analysis (PCA), probably the most popular dimension reduction technique. First, we very briefly review classical PCA, then we proceed with a selection of proposals for robust PCA: these include simple methods performing PCA based on robust covariance estimates, spherical PCA and projection pursuit to deal with high dimensional problems. A section is devoted to techniques and advantages of outlier identification after dimension reduction. The chapter is concluded with the detailed development of three examples. Chapter 4 is devoted to *sparse* PCA, that is, (robust) PCA techniques which do not make use of all measured covariates. Special attention is given to the choice of the degree of sparsity of the robust solution. Two examples conclude the chapter. Chapter 5 deals with canonical correlation analysis. First, the classical method is reviewed and, then, some methods for robust CCA are illustrated, including CCA based on robust covariance estimation, spherical CCA, and projection pursuit. The first part of the book is closed by a brief chapter covering factor analysis and its robust counterparts. Two examples aid the dissertation in both chapters.

The second part of the book is devoted to cluster and discriminant analysis. By finding groups in the data we are actually performing *sample reduction*, that is, we summarize information by reducing the number of observations we have. It should be noted that while in the first part we devote one chapter to each technique, beginning with a review of the non-robust approach. The second part follows a different scheme. In Chapter 7, we review non-robust distance and model-based approaches to cluster analysis, giving the general set up and main ideas in sample reduction. In Chapter 8, we then introduce one of the first approaches to robust clustering, that is, partitioning around medoids, and then proceed to more recent procedures based on the concepts of trimming and snipping. Chapter 8 is then limited to distance-based methods. In Chapter 9, we complement the previous chapter describing extensions to model-based clustering. The second part of the book is concluded with a chapter on simultaneous clustering of rows and columns of a data matrix and a brief chapter on discriminant analysis.

It is important to underline that we by no means cover all, or even the majority, of robust procedures in data reduction. An impressive number of approaches have appeared in the literature. This book is not an attempt to systematically review all those, but rather our personal selection of the ones we believe are simpler to understand and/or to use in practice (and of course, effective). Recall that the scope of this book is not to provide deep insights into robust statistics, but rather to encourage use of robust data reduction in practical applications.

Several libraries in the R software (R Development Core Team, 2012) have been implemented to perform robust analyses. We will try to illustrate these functions both in the main chapters and in a separate appendix. For robust techniques not included in any R package, we make our own code available in the publisher's website dedicated to this book. Routines to perform robust analyses are also available from other mathematical and statistical software. The FSDA MATLAB Toolbox described in Riani *et al.* (2012) is particularly worth mentioning.

We try to make an extensive use of real data examples of increasing complexity. The examples in the first two chapters are used mostly for a simple and practical illustration of the main ideas on robustness. The examples in the rest of the book are aimed at bringing evidence in favor of the practical use of robust methods in data reduction. Also the data and R code for developing most of the examples are made available and can be downloaded freely from the book's web page.

Authors

Alessio Farcomeni is assistant professor, Department of Public Health and Infectious Diseases, Sapienza - University of Rome. He has interests in robust statistics, longitudinal models, categorical data analysis, cluster analysis, and multiple testing. He also intensively collaborates in clinical, ecological and econometric research.

Luca Greco is assistant professor, Department of Law, Economics, Management and Quantitative Methods, University of Sannio. He has interests in robust statistics, likelihood asymptotics, pseudo-likelihood functions, and skew elliptical distributions.

List of Figures

List of Tables

List of Examples and R illustrations

Symbol Description

n	Sample size	$\bar{X}$	k by p matrix of centroids
p	Number of variables	z	Binary vector indicating selected and discarded rows of X
X	n by p data matrix		
F	Model assumed		
F_ϵ	Contaminated model	Z	Binary matrix indicating selected and discarded entries of X
k	Number of clusters		
q	Number of components		
S	p by p covariance matrix	π	Vector of cluster weights
μ	Population mean	I_p	Identity matrix of size p
$\bar{x}$	Sample mean	$d(\cdot, \cdot)$	Generic distance function
$\hat{\mu}$	Location estimate	$l(\cdot)$	Loss or likelihood function
σ^2	Population Variance	d_{ic}^2	Euclidean or Mahalanobis distance
s	Sample standard deviation		
$\hat{\sigma}$	Scale estimate	$\Phi(x)$	Standard normal distribution function
ψ	Estimating function		
ψ_H	Huber estimating function	$\Phi_p(x, \mu, \sigma)$	Multivariate normal distribution function
ψ_T	Tukey estimating function		
w_i	Individual weight	$\phi(x)$	Standard normal density function
Σ	Population covariance matrix		
		$\phi_p(x, \mu, \Sigma)$	Multivariate normal density function
λ	Eigenvalue	$g(\cdot)$	Arbitrary (usually contaminating) density
Λ	Diagonal matrix of eigenvalues		
ν	Eigenvector	ε	Trimming, snipping and contamination level
V	Matrix of eigenvectors		
Y	n by p matrix of principal components	ε^*	Asymptotic breakdown point
ρ	Correlation coefficient	$\varepsilon^{(i)}$	Individual breakdown point
a_{ij}	Loadings		
f_s	Factors	$\varepsilon^{(c)}$	Cell breakdown point
h_j	Communality (in factor analysis)	γ^*	Gross error sensitivity
		ASV	Asymptotic variance
Ψ	Uniqueness (in factor analysis)	IF	Influence function
		$c(p, \varepsilon)$	Consistency factor for the MCD
U	n by k binary matrix with cluster labels	$const$	Generic constant

1

Introduction and Overview

Robust statistics is an extension of classical statistics that specifically takes into account the fact that models only provide an approximation to the *true* underlying random mechanism that generates the data. Model assumptions are almost never exactly satisfied in practice. A fraction of the observations can exhibit patterns not shared by the bulk of the data and therefore be *outliers*. The occurrence of departures from model assumptions by atypical values may have unexpected and deleterious effects on the outcomes of the analysis. These issues are interwined: an inappropriate model can be the reason of several data anomalies, and many outlying observations may suggest that the model is not adequate. A fundamental concept is that outliers are such only with respect to a certain model. Under the model, these observations are very unlikely or even impossible.

We begin to illustrate concepts in robust statistics with a very simple univariate scenario.

1.1 Example (Simulated univariate Gaussian) *Consider a sample x of size $n = 30$ from a Gaussian distribution with expected value $\mu = 25$ and standard deviation $\sigma = 5$. The sample can be generated with the R function* `rnorm(30,25,5)`. *Using the* `round` *function, we obtain the following values:*

28.69	23.81	27.27	26.49	26.95	22.24
29.86	24.59	27.25	26.16	22.76	25.60
23.75	16.54	23.58	24.66	25.43	24.14
22.74	31.08	26.06	22.66	29.91	15.34
28.85	23.88	24.03	31.93	22.15	27.95

The sample mean and standard deviation are $\bar{x} = 25.21$ and $s = 3.68$, obtained with the R functions `mean` *and* `sd`, *respectively. Now let us suppose that just one anomalous value corrupts the original sample: for illustrative purposes, we replace the maximum value 31.93 with a more extreme 61.93 (that, for instance, may be due to a transmission or typing error). The sample mean and standard deviation become 26.21 and 7.58: both summaries have increased. The damaging effect of this outlier on inferential procedures is more evident when computing 95% (exact) confidence intervals for the parameter μ: based on the original sample it is $(23.84, 26.58)$, whereas it turns into $(23.38, 29.04)$ because of the presence of the single outlier. The anomalous value doubled the interval's length. If we instead use the sample median, we notice that the* *outlier does not affect its value, which corresponds to 25.05 before and after its inclusion. For that matter, we could arbitrarily increase any value above 25.04 or decrease any value below 25.05 and still obtain the same median. The same applies to the Median of Absolute Deviations (MAD), which is the median of the absolute distances from the median (multiplied by 1.4826 to approximate the standard deviation), that is equal to 3.28 both in the original and contaminated sample. The median and the MAD can be obtained in R with the functions* `median` *and* `mad`, *respectively. A robust (asymptotic) confidence interval for μ based on the couple (median, MAD) is $(23.57, 26.52)$ in both the original and contaminated sample.*

A summary of this toy example is that the median and the MAD are resistant to outliers and inference based on them is robust with respect to departures of the data from the specified model. On the other hand, when the model is correct and there are no outliers, there is a slight loss of efficiency which is testified by the increased length of the confidence interval based on the robust summaries.

This book is mainly concerned with the analysis of multivariate data and the nature and effect of multivariate outliers. We provide a second toy example in which the sampling model is a *bivariate* Gaussian distribution.

1.2 Example (Simulated bivariate Gaussian) *Consider a sample of size $n = 20$ drawn from a bivariate Gaussian distribution $N_2(\mu, \Sigma)$ with null mean vector $\mu = (0, 0)^{\mathrm{T}}$, unit marginal variances and correlation 0.5, stored in the $n \times 2$ data matrix X. The data can be generated by using the function* `rmvnorm` *from library* `mvtnorm`, *according to the following code*

```
n<-20
Sigma<-matrix(c(1,0.5,0.5,1),2,2)
X<-rmvnorm(n,sigma=S)
```

and our random draw is available from the supplementary material of this book for reproducibility. Figure 1.1 shows 90% level tolerance ellipses of the form

$$\left\{ \mu = (\mu_1, \mu_2)^{\mathrm{T}} : (\hat{\mu} - \mu)^{\mathrm{T}} \hat{\Sigma}^{-1} (\hat{\mu} - \mu) \le q_{.90} \right\},$$

where $q_{.90}$ is a 90%-quantile of a reference distribution to be detailed in Section 2.4, $\hat{\mu}$ is an estimate of the true mean μ and $\hat{\Sigma}$ is an estimate of the true covariance matrix Σ. The last two can either be the sample mean vector $\bar{x} = (\bar{x}_1, \bar{x}_2)^{\mathrm{T}}$ and the sample covariance matrix S, or robust estimates of multivariate location and covariance. In this example we will use the reweighted MCD (Minimum Covariance Determinant), to be defined in Section 2.3.4. Estimates are obtained for the original sample (in the top left corner of Figure 1.1) and when two outliers (denoted by $+$) are added in different locations. Drawing ellipses is possible through the function `ellipse` *after loading the R package* `ellipse`. *The robust fit is close to the classical one for the original sample. In all contaminated scenarios, the presence of outliers has a severe*

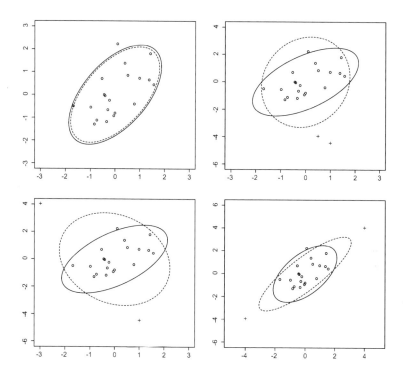

FIGURE 1.1
Classical (dotted line) and robust (solid line) tolerance ellipses for the original sample and when two outliers (indicated by a +) are added.

effect on the classical estimates: the tolerance ellipse is always shifted towards them. On the other hand, the shape and location of the robust tolerance ellipse are almost unchanged: the robust fit is not affected by contamination. It shall be noted here that multivariate outliers, as will be detailed later, may be of different nature. In the top right panel, the two outliers are clearly far from the bulk of the data, but they are atypical only with respect to one dimension. In the bottom left panel, only one outlier is anomalous in both dimensions; in the last panel, both are anomalous in both dimensions.

An ideal robust procedure will behave like the ones we saw so far: it will be resistant to contamination in the data and have a tolerable efficiency loss with respect to the classical procedure when model assumptions are satisfied.

Nowadays it is well acknowledged that common tools for data analysis shall not be used in presence of contamination. Simply ignoring the fact that some observations are unusual according to the assumed model leads to biased estimates and completely wrong conclusions. Furthermore, screening for contamination and estimation should be performed simultaneously. The common approach of screening for outliers, maybe by visual inspection, and then discard them before data analysis is flawed in many ways. First of all, it may be extremely hard to identify outliers when the dimensionality is large, hence contamination may still be present after visual inspection. Additionally, a problem with some naive outlier identification procedures is that they are not based on robust procedures, and therefore they can lead to masking and swamping, that is, to misclassification of outliers. There is masking whenever an outlier is not identified correctly, while there is swamping when a genuine observation is flagged as aberrant. Secondly, when inference is based on a subjectively chosen subset of the available data, we may be underestimating the impact of model misspecification. Uncertainty in outlier labeling is ignored, that is, labels are treated as known when they actually are not. Third, in some cases it is more efficient to downweight rather than discard suspicious observations. Finally, and most importantly, outliers are actually model dependent and, therefore, procedure dependent. An observation which is deleterious for a procedure may instead be useful for another one. This is a consequence of the fact that we define outliers with respect to modeling assumptions. When we have a different objective of the analysis we may also be using different assumptions, therefore obtaining a different set of outliers for the same data set.

1.1 What is contamination?

Contamination in the data may be meant as the presence and overlap of different random mechanisms: one is assumed to generate the majority of the data, the other is responsible for the occurrence of outliers. Even if we are able to specify a model to fit the majority of the data, it is not obvious how to choose directions of departures from model assumptions. In this sense, we shall employ procedures that are robust to contamination arising from models which are as general as possible. In this book we will interchangeably use the terms *contamination* and *outliers* to indicate a general departure from model assumptions. Outliers in the data are supposed to be unexpected values, whose random generating process can not be unveiled by the analysis. On the other hand, after outliers are identified, it can be useful and interesting to classify them somehow. In this section we give a brief and non-exhaustive account of some of the kinds of contamination we can encounter in real data applications. Definition and classification of multivariate outliers has been a lively area of research in the past decades. A general account can be found in Barnett and Lewis (1994), Davies and Gather (1993), Markatou *et al.* (1998), and Gather *et al.* (2003) who define outliers as values sampled in regions receiving small probability under the assumed model. The latter is an encompassing definition, which can be specialized in our opinion with the following incomplete classification, suited to the multivariate framework:

- *extreme values*

- *influential outliers* or *leverage points*

- *inliers*

- *bridge points*

The most important distinction is between extreme and influential outliers. The first are simply unusually large or small values with respect to one or more dimensions (sometimes also referred as *gross outliers*). They follow the general trend and in practical terms do not give controversial evidence. As an example, consider the situation of two variables with a strong positive correlation, in which a data point shows a large value for both. Influential outliers are instead data points which do not follow the pattern shared by the remaining majority of the data. As an example, consider again the situation of two variables with a strong positive correlation, in which a data point with a large value for one shows a small value (or even a value close to the average) for the other. Influential outliers are surprising and can lead to discover rare subpopulations or to unveil confounding which was not previously considered. In simple words, influential outliers are deleterious if ignored but potentially hypothesis generating once identified. Inliers (Akkaya and Tiku, 2005, 2008) are corrupted values in the center of the distribution. An influential outlier can be an inlier if it is not extreme under any dimension, for instance. Finally, bridge points are values placed in a low density region surrounded by two or more high density regions. These outliers are particularly important in cluster analysis, where high density regions are the clusters and bridge points are difficult to attribute to any of the identified groups.

We further distinguish two complementary kinds of outliers: *structural outliers* (Tukey, 1962; Huber, 1964; Gallegos and Ritter, 2005) and *component-wise outliers* (Alqallaf *et al.*, 2009; Maronna and Yohai, 2008). In the first case a small fraction of observations are outliers arising from their own population, different from that of the rest of the data. Component-wise contamination occurs, instead, when each dimension of an observation can be separately contaminated. In this setting contamination arises more often during the measurement rather than the sampling phase. Observations are sampled from the specified model, but the measurement of each dimension can be (independently) contaminated. To clarify the idea, suppose we are measuring gene expression levels repeatedly on independent slides. We have structural outliers when one or more genes arise from a different pathway than the one under study, while we have component-wise contamination due to contamination within each slide. It is not difficult to imagine many situations in which component-wise contamination can arise. Most robust procedures treat multivariate outliers similarly regardless of whether they are structural or component-wise outliers, with few exceptions (Maronna and Yohai, 2008; Alqallaf *et al.*, 2009; Farcomeni, 2014a,b; Agostinelli *et al.*, 2014; Farcomeni, 2015).

1.3 Example (Birth and death rates in Italy) *We illustrate the different kinds of anomalous measurements in a simple example on natality and mortality rates in Italy. The natality and mortality rates in 1997 were recorded for the 20 Italian regions. Hence in this example $n = 20$ and $p = 2$. A scatterplot is shown in Figure 1.2, where a clear trend is apparent: as the birthrate increases, the death rate decreases proportionally. This is a well known effect in demography, and is related to population aging and similar considerations: a region with a large birthrate has a certain proportion of young couples and young babies which is basically replaced by elders in regions with a large death rate. The decreasing trend is seen since the latter are more likely to experience the event than the former. In Figure 1.2 we visually identify two extreme values and one influential outlier: in the top left corner we find Liguria, which has the largest proportion of elderly in Italy, and at the lower right corner of the data cloud we have Campania. Finally, we have a clear influential outlier, Sardegna. Given the birth rate, Sardegna would be expected to have a much larger death rate. This is not surprising as it is well known that Sardinians live longer and are less prone than the rest of Italians, due to genetic factors, to various neuroendocrine diseases (e.g., Capalbo et al. (2012) and references therein). The objective of a robust analysis of these data is twofold: first, obtain a fit for the bulk of the data, which is not influenced by Sardegna. Then, use formal outlier identification procedures to classify data points and confirm what Figure 1.2 has suggested. Is Sardegna a leverage point? Can we formally conclude that Liguria and Campania are atypical with respect to the other Italian regions?*

1.2   Evaluating robustness

A robust statistic (more generally a robust method) is expected to be defined so that outliers have a *bounded* effect on it, and to have large efficiency at the assumed parametric model (which means in absence of contamination). Furthermore, it is supposed to resist the largest possible rate of contamination before it *breaks down* and have reasonable *maximum bias* for every rate of contamination it can tolerate. These concepts are better defined in this section.

When these robustness requirements are not fulfilled, inference can be unreliable under contamination. These properties can be investigated by means of formal tools like the influence function (IF) , the breakdown point (BP) and the maximum bias (MB). These can be considered as the main measures of robustness.

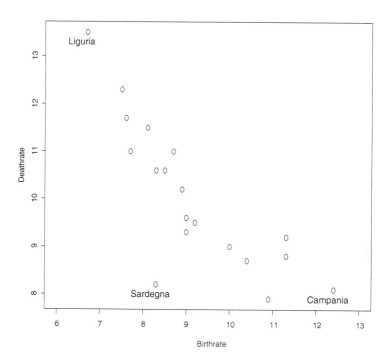

FIGURE 1.2
Italian demographics data.

1.2.1   Consistency

Let X denote a data set of size n. This will be a vector with n elements in the univariate case $(p = 1)$, or a matrix with n rows in the multivariate setting $(p > 1)$. Let F denote the assumed sampling model, that is a member of a parametric family of distributions

$$\mathcal{F} = \{F|F = F(X; \theta), \theta \in \Theta, X \in \mathcal{X}\} \ ,$$

where θ is a general (vector) parameter, Θ is the parametric space and $\mathcal{X}$ the sample space.

A statistic $T(X)$ can be expressed as a functional of the form $T(F_n)$, where F_n is the empirical distribution function

$$F_n(x) = \frac{1}{n} \sum_{i=1}^{n} I\{X_i \leq x\} \ ,$$

where $I(E)$ is the indicator function of the event E. This generic definition en-

compasses statistics estimating location, scale, covariance, eigenvalues, eigenvectors or factor loadings, for instance. A well-known result in asymptotic
statistics is that the functional $T(F_n)$ is such that

$$T(F_n) \overset{p}{\to} T(F)$$

as the sample size grows, that is the statistical functional *converges in probability* to a value $T(F)$, that is the *asymptotic value* of $T(F_n)$ at the model F.
In words, the sample statistic converges to the true population statistic.

Definition 1.1 *Let F be an element of the parametric family of distributions
$\mathcal{F}$. The statistic $T(F_n)$ is Fisher consistent at the model F if $T(F) = \theta$, $\forall$
$\theta \in \Theta$, $F \in \mathcal{F}$*

Fisher consistency means that the statistic $T(F_n)$ (an estimator $\hat{\theta} = T(F_n)$,
for instance) converges to the parameter θ as soon as F_n converges to F
(Hampel *et al.*, 1986; Maronna *et al.*, 2006; Huber and Ronchetti, 2009).
   What happens when a fraction of the data is contaminated? Where does
$T(F_n)$ actually converge? To what extent is inference based on $T(F_n)$ robust? What is the largest amount of contamination that can be tolerated
before inference based on $T(F_n)$ becomes too liberal? In order to answer these
questions, we need to investigate the behavior of $T(F)$ when the model F
is contaminated, i.e., when the data are not generated from F, but from a
model close to it. A simple way to define a neighborhood of the model F
is through the ϵ-neighborhood or Tukey-Huber contamination model (Tukey,
1962; Huber, 1964):

$$\mathcal{P}(F, \epsilon) = \{F_\epsilon | F_\epsilon = (1 - \epsilon)F(X; \theta) + \epsilon G(X), \theta \in \Theta, X \in \mathcal{X}\} , \qquad (1.1)$$

where G is an arbitrary distribution from which outliers are drawn. In words,
$(1 - \varepsilon)\%$ of the data arises from F, and the remaining small fraction ε are
arbitrary structural outliers from G.

1.2.2   Local robustness: the influence function

The influence function (Hampel, 1974; Huber, 1981) describes the effect of
departures from the specified parametric model F within a neighborhood
$\mathcal{P}(F, \epsilon)$. The influence function is given by the the relative change in $T(F)$,
such as a point estimate or the level/power of a test statistic, caused by a
small proportion ε of outliers all equal to an arbitrary value $x \in \mathbb{R}^p$.

Definition 1.2 *Let $F_\epsilon = (1 - \epsilon)F + \epsilon\delta_x$, where δ_x is a point mass contamination at x. The IF for an infinitesimal point mass contamination ϵ at location
x at the model F is*

$$IF(x; T, F) = \lim_{\epsilon \to 0} \frac{T(F_\epsilon) - T(F)}{\epsilon} = \frac{\partial}{\partial \epsilon}T(F_\epsilon)|_{\epsilon=0} . \qquad (1.2)$$

The quantity $\epsilon IF(x; T, F)$ gives an approximation of the bias of $T(F_\epsilon)$ determined by the *infinitesimal* contamination. A bounded influence function means that the relative change in the functional is never arbitrarily large. Hence, if $IF(x; T, F)$ is bounded, then a point mass contamination located at x can not be too deleterious for the statistical properties of T at the model F. If $IF(x, T, F)$ is bounded for all x, we have formally shown *local robustness* of T.

The relative effect of a point mass contamination as measured by the IF can be summarized by the *gross error sensitivity*

$$\gamma^* = \sup_{x \in \mathbb{R}^p} ||IF(x; T, F)|| . \tag{1.3}$$

A bounded γ^* identifies a (bias-) robust statistic.

A second index that can be obtained from the IF is aimed at measuring the effects of small changes in the data, as those arising from rounding and grouping, for instance. Let the observation x be replaced by a *new* value y. The effect of shifting can be measured by the *local shift sensitivity*

$$\lambda^* = \sup_{x \neq y, x, y \in \mathbb{R}^p} \frac{||IF(y; T, F) - IF(x; T, F)||}{||y - x||} . \tag{1.4}$$

If $y = x + \epsilon_x$, with $\epsilon_x \to 0$, then the local shift sensitivity corresponds to the slope of the IF at the point x.

A third summary value directly related to the IF is the *rejection point*, that identifies points where the IF vanishes and therefore contamination has no effect. See Hampel *et al.* (1986) and Huber and Ronchetti (2009) for an extensive discussion.

An empirical tool to measure the effect caused by the occurrence of one outlier in the sample is the sensitivity curve (SC). The SC describes how a statistic changes as a function of one outlying observation that is added to the original sample. It can be used to investigate the robustness properties of the statistic by means of empirical considerations and graphical displays.

Definition 1.3 *Let us consider a statistic $T(X)$, a sample x and a new observation x_0. The sensitivity curve is*

$$SC(x_0) = T(x, x_0) - T(x) , \tag{1.5}$$

as a function of x_0.

The influence function (1.2) can be interpreted as the limit version of the sensitivity curve (1.5). If x_0 is added to the sample, there is by definition a contamination of size $\epsilon_n = 1/(n+1)$, and for large n the curve

$$\frac{T(x_1, x_2, \ldots, x_n, x_0) - T(x_1, x_2, \ldots, x_n)}{\epsilon_n}$$

approaches $IF(x; T, F)$, since ϵ_n becomes infinitesimal as n grows.

The influence function is also a fundamental tool to assess the asymptotic behavior of a statistic and, consequently, its efficiency at the assumed model.

Proposition 1.1 *Let $T(F_n)$ be Fisher consistent for $\theta \in \Theta$ at the model F, and $IF(x; T, F)$ be the corresponding influence function. Then*

$$\sqrt{n}\,(T(F_n) - T(F)) \xrightarrow{d} N_p(0, ASV) \qquad (1.6)$$

with

$$ASV = ASV(T, F) = \int IF(x; T, F) IF(x; T, F)^{\mathrm{T}}\, dF(x)\,. \qquad (1.7)$$

The asymptotic relative efficiency (ARE) of $T(F_n)$ at the model F can be evaluated by the ratio of the asymptotic variance of the maximum likelihood estimate (MLE) to its asymptotic variance obtained from (1.7). The asymptotic relative efficiency of a robust estimator will always be strictly smaller than the unity, but we will in general be able to set lower bounds so to guarantee a certain efficency (e.g., 90% or 95%).

1.2.3   Global robustness: the breakdown point

The breakdown point of a statistic $T(F_n)$ is the largest fraction of the data (amount of contamination) that can be arbitrarily replaced by outliers while $T(F_n)$ is bounded. Up to a certain rate of arbitrary contamination a robust statistic is expected to be bounded away from the boundary of the parameter space. This does not happen for instance with the sample mean: a single infinite observation leads to an infinite mean, regardless of the sample size. The largest amount of contamination that is tolerable (at least asymptotically) by many robust statistics is 50%, while many classical estimates (like the sample mean) can bear no contamination at all.

There are different possible definitions of breakdown point, based on asymptotic arguments (Hampel, 1971), on the finite sample (Hodges, 1967; Donoho and Huber, 1983) behavior of a statistic, or on the nature of contamination.

Definition 1.4 *The asymptotic breakdown point (BP) of a statistical functional $T(F)$ is the largest rate of contamination ϵ such that $T(F_\epsilon)$ is bounded and bounded away from the boundary of the parameter space, i.e.*

$$\varepsilon^* = \max\left\{\varepsilon : T(F_\epsilon) \in K \subset \Theta\right\},$$

where K is a bounded and closed set that does not contain the boundary points of the parametric space.

Definition 1.5 *Let $X_r \in \mathcal{X}_r$, where $\mathcal{X}_r$ is the collection of all data sets X_r of size n having $(n - r)$ elements (that is entire rows) in common with the original data X. Let K be any bounded and closed set that does not contain the boundary points of the parameter space. The finite sample breakdown point is*

$$\varepsilon^{(i)} = \max\left\{\frac{r}{n} : \sup_{\mathcal{X}_r} \|T(X) - T(X_r)\| \in K\right\}.$$

In words, $\varepsilon^{(i)}$ measures the largest fraction of arbitrary structural outliers we can include without obtaining a diverging $T(X)$. In a similar fashion and simply speaking, ε^* is the breakdown value of a procedure as the sample size grows to infinity. If a breakdown value is infinitesimal, we can formally declare that the procedure is not robust.

The breakdown points described so far correspond to replacement of entire rows of the data matrix with arbitrary values. These are formally regarded as *individual breakdown points* (e.g., Gallegos and Ritter (2005)). We now deal with another kind of corruption, particularly relevant in the multivariate case, that is brought about by component-wise contamination. The corresponding breakdown point is obtained by replacing isolated entries of the data matrix X, rather than entire rows. A formal definition of *cell breakdown point* was given in Farcomeni (2009b, 2014b). It can be formalized as follows:

Definition 1.6 *Let $X_c \in \mathcal{X}_c$, where $\mathcal{X}_c$ is the collection of $n \times p$ data matrices X_c having $np - c$ entries in common with the sample X. Let K be any bounded and closed set that does not contain the boundary points of the parameter space. The cell breakdown point is*

$$\varepsilon^{(c)} = \max \left\{ \frac{c}{np} : \sup_{X_c} \|T(X) - T(X_c)\| \in K \right\}$$

For a definition of breakdown points when estimators have bounded support, refer to Genton and Lucas (2003). In Definition 1.5 and Definition 1.6 we have ignored the dependence of $\varepsilon^{(i)}$ and $\varepsilon^{(c)}$ on the data set X, implicitly taking the supremum over all possible data sets. These are the so called *universal* breakdown point, which are very rigorous and describe a worst case scenario in which the clean data itself is completely arbitrary and can be very far from the assumed model. The universal breakdown point is not particularly relevant in cluster analysis, where it amounts to evaluating the ability of obtaining k groups also for data sets which could hardly be close to such a structure. To tackle this issue Gallegos and Ritter (2005) have introduced the concept of *restricted* breakdown point, which is the (minimal) breakdown point obtained for data sets within a certain class (for instance the class of data sets which are well separated into k groups).

We conclude noting that if the gross error sensitivity of an estimator is not bounded, the breakdown point (however defined) is also consequently infinitesimal.

1.2.4 Global robustness: the maximum bias

The IF gives information about the asymptotic bias of a statistic under an infinitesimal contamination. The BP deals with the largest amount of contamination a statistic can withstand before becoming degenerate. They are both uninformative about the actual magnitude of bias that can be expected with a certain rate of contamination. The (asymptotic) bias of a statistical functional can be analyzed by means of the maximum bias. See Rousseeuw (1998) for a summary of the topic.

Definition 1.7 *The maximum bias of the statistical functional $T(F)$ on the neighborhood $\mathcal{P}(F, \epsilon)$ is measured by*

$$MB(T, F, \epsilon) = \sup_{F_\epsilon \in \mathcal{P}(F,\epsilon)} ||T(F_\epsilon) - T(F)|| . \qquad (1.8)$$

The maximum bias is related to the asymptotic breakdown point by the relationship

$$\varepsilon^* = \max \{\epsilon : MB(T, F, \epsilon) < \infty\} .$$

For a given contaminating distribution G, by letting ϵ vary between zero and ε^* we obtain the *maxbias curve*. Additionally, if the function (1.8) is differentiable at $\epsilon = 0$, its first derivative is the *contamination sensitivity*. The reader is pointed to Yohai and Maronna (1990); Martin and Zamar (1993); Croux and Haesbroeck (2001, 2002) for a more detailed discussion.

1.3   What is data reduction?

Data reduction is the process of summarizing the data matrix by aggregating information. Smaller information nuggets are obtained by aggregating measurements, so that information is manageable. An informal constraint is that loss of information should be as small as possible, that is, use of the reduced data matrix produces similar conclusions compared to those that would be suggested by the original data.

Given that (most) data matrices are two-way, with cases recorded in rows and variables in columns, it is natural to think about data reduction as what is obtained by reducing the number of variables, the number of rows, or both. In the first case we perform *dimension reduction*, in the second *sample reduction*. Note that the two procedures may be performed simultaneously (Law *et al.*, 2004; Van Mechelen *et al.*, 2004; Madeira and Oliveira, 2004; Farcomeni, 2009b, e.g.).

Data reduction is possible as measurements can be very similar. From the variables perspective, highly correlated measurements give approximately the same information. In the extreme case in which variables are approximately linearly dependent, we can reduce the dimensionality from p to 1 with negligible loss of information. For all practical purposes, a strong dependence corresponds to a good possibility of compression. In general, the p measurements will be replaced by $q << p$ *scores*, which are obtained as (linear) functions of the old variables. From the observations perspective, in the extreme case in which subjects give approximately equal measurements we can reduce the sample size from n to 1 with negligible loss of information. For this reason, a low variability corresponds to a good possibility of compression. In general, the sample of size n will be reduced to a sample of size $k << n$ *clusters*, whose profiles will replace the old observations. A cluster profile can be simply defined through an opportune summary statistic (e.g., mean or median). Note that while it is common to summarize the loss of information due to dimension reduction, the same could in principle be done with sample reduction but it is seldom seen in reality.

There can be many goals in mind when performing data reduction. Among those, we mention compression, enhancement of interpretability, use in further analyses, ranking, possibility to plot (perceptual mapping), summarize (that is, obtain estimates stratified by an unknown or even known labeling). We will try to stress most of these tasks when dealing with the reviewed procedures and in real data examples. Note that for many of the goals of data reduction to be meaningful there is actually no need of starting with a large data matrix in the first place.

1.3.1 Dimension reduction

Dimension reduction is a desired task in all those cases in which one wants to extract meaningful patterns and associations from a large dimensional data set, and identify the most important features across all variables. A first aim behind dimension reduction is purely descriptive, leading to explore the interdependence among variables. The assessment of the most (and less) important variables in terms of total variability is usually realized by obtaining *new* variables as *linear combinations* of the original ones. This procedure helps the researcher in identifying the most important relationships in the data. The new variables are computed so to have the largest possible variance. The larger the *explained* variance of the new variables, the more information is retained within the data reduction procedure. An effective dimension reduction is achieved when few ($q << p$) linear combinations are able to store a very large part of the original total variability. Additionally, the largest (in absolute value) coefficients of the linear combination identify the most relevant variables and unveil otherwise unobserved relationships.

The most popular linear technique for dimensionality reduction is *principal component analysis* (PCA). PCA provides a strategy to obtain a simpler representation of the data, for feature extraction and compressed storing of information. PCA is based on *projecting* the p dimensional data onto a lower (one, two or three, for instance) dimensional space. The first principal component can often be used as an optimal summary of all measurements, useful for instance for ranking subjects. The first two or three principal components allow us to produce a graphical display of the p dimensional data set.

Few components can often be sufficient to describe economical or physical phenomena, which are otherwise measured by a large number of characteristics. Dimension reduction accomplished through few linear combinations is particularly appealing in the analysis of high dimensional data sets, as those arising from gene expression studies, where thousands of gene expression levels are measured in a limited number of experiments. In chemometrics, infrared spectra are composed by thousands of wavelengths which can often be successfully summarized in few dimensions. Detection of the most important genes or bands can undoubtedly benefit from the *extraction* of a small number of components accounting for a large percentage of variability and characterized by a relatively small number of large coefficients. A simple example is given in Farcomeni *et al.* (2008). In this book we also deal with the problem of *sparseness* of principal components. Sparseness corresponds to the situation in which each principal component is a linear combination of only a subset of variables. Sparseness substantially enhances interpretability and reliability of dimension reduction.

1.3.2   Sample reduction

When we believe that our observations are separated within unobserved groups, we can perform sample reduction. In order to reduce the sample size we usually obtain *new* observations which are summaries (that is, cluster profiles or *centroids*) of groups of similar observations. More precisely, sample reduction is performed by *partitioning* a data set into subsets, so that the units within each subset are similar.

Cluster analysis attaches a label, from 1 to k, to each of the observations. Observations with the same label are in the same group and are therefore similar with respect to the measurements used by the procedure. An important step after cluster analysis is that of interpretation and evaluation of clustering. We will focus in this book on *crisp* clustering procedures, as opposed to *fuzzy* clustering. In the former, subjects are assigned at most to one group, while in the latter a continuous degree of membership is defined, and subjects may be actually assigned to all groups with a fractional degree of membership.

Two observations are assigned to the same cluster when they are close. It very often happens that proximity is defined according to some distance measure. The choice of a distance affects the cluster shape, that is, how clusters are defined. This is something that is important to keep in mind: the results of sample reduction may be heavily dependent on the cluster shape, and this choice is often intrinsic, as it is a consequence of the proximity measure chosen. Clusters could be spherical, linear, elliptical, etc. In this book we will mostly use the Euclidian distances (therefore having spherical clusters) or the Mahalanobis distances (therefore having elliptical clusters). It is important to underline that the optimal number of clusters heavily depends on the cluster shape. More flexible shapes (e.g., elliptical rather than simply spherical) often lead to a lower optimal number of clusters.

There can be different goals when performing cluster analysis. In *exploratory* cluster analysis, sample reduction is performed basically as a descriptive device. Few clusters of large dimensionality can be obtained for targeting few policies (e.g., marketing policies) to large groups of subjects. Cluster analysis is often used in ecological modeling as a preprocessing tool before building maps of potential or observed species distributions (e.g., De Sanctis *et al.* (2013), Attorre *et al.* (2014)). In *confirmatory* cluster analysis, a known labeling is ignored and data are clustered to see if they are well separated with respect to the known labeling. In practice, the cluster labels are compared for agreement with the known labels, in order to verify their discriminatory ability. This is useful for instance when labels should identify well separated groups of individuals, such as discrete risk scores in clinical medicine. Risk scores are often built based on clinical and demographic predictors, and their validation with respect to the outcomes is often of interest. We will often perform confirmatory cluster analysis in our examples, since it is a useful means of illustrating the performance of clustering methods on real data.

The performance of most clustering algorithms, and in particular of robust clustering algorithms is usually heavily influenced by the class of data sets at hand. Groups can be well separated, connected by bridge points, or overlapping. The output of any clustering algorithm shall always be carefully evaluated. All clustering algorithms will produce clusters, regardless of whether the data contain them or not. It is a further task of the user to determine if clusters are well separated (if not, k is too large) and if they are homogeneous enough (if not, k is too small). There is anyway no optimality criterion in cluster analysis which can go beyond the final aim of the application.

For most of the book we will assume data cluster labels are unknown. This framework is referred to as cluster analysis, or unsupervised learning.

A notable exception is in Chapter 11, where we will assume that group labels are known (that is, that we want to reduce the sample according to a known categorical measurement). This framework is usually referred to as classification, or supervised learning. When labels are known we can simply perform sample reduction with respect to a known labeling, or build a classification rule for future observations. In the first case, one can compute stratified statistics (e.g., cluster means, medians, standard deviations) and use those for further analysis or conclusions. In the second case one can use a classification technique. In this book we will review only one classification technique, (robust) discriminant analysis, but we mention a large number of classification techniques are available and point the reader for instance to Hastie *et al.* (2009). For non-robust versions of discriminant analysis refer for instance to McLachlan (1992) or to Huberty (1994).

1.4 An overview of robust dimension reduction

Dimension reduction techniques are aimed at explaining the covariance structure of the data by means of a small number of uncorrelated linear combinations of the original variables. When outliers are likely to occur, and particularly with a large number of observations and/or variables, these new variables should be based on some robust measure of variability. A common approach to robust dimension reduction is based on the use of a robust estimate of the covariance matrix. This is achieved for instance by downweighting those observations that are mostly suspicious. Several weighting schemes have been proposed: some are based on continuous weights (Salibian-Barrera *et al.*, 2006), some on hard rejection rules based on $0 - 1$ weights (Croux and Haesbroeck, 2000). In the first case, observations contribute to the analysis according to their degree of outlyingness, in the second case anomalous values are trimmed and removed from the study. Weights are always determined according to the *distance* of each multivariate data point from the robust fit (Maronna *et al.*, 2006; Huber and Ronchetti, 2009). Continuous weights are inversely proportional to this distance, while discrete weights are obtained after thresholding. A different strategy relies on the use of robust measures of scale after multivariate data have been projected onto univariate directions (Croux *et al.*, 2007). The search of appropriate univariate directions is the core of projection pursuit techniques.

In order to provide a brief overview of robust dimension reduction we use data concerning a study of 16 colonies of the butterfly *Euphydryas edhita* in California and Oregon (McKechnie *et al.* (1975), Manly (2005)). In particular, we focus on four environmental variables (altitude, annual precipitation, minimum and maximum temperature): univariate and pairwise graphical summaries of the data, along with the correlations, are displayed in Figure 1.3. The figure has been obtained using the R function `pairs`, with some additional options. We observe correlations far from zero, but also points that remarkably deviate from the trend in some pairwise panels. A closer look at the data lead us to detect two colonies, namely `GH` and `GL`, characterized by very high altitude and low temperatures.

It is worthwhile to investigate if and how these two colonies influence the correlations, and what are the consequences on dimensionality reduction through PCA.

First, we compute robust correlations based on the reweighted MCD and MM estimator, that will be discussed later in the book. As could be expected, values in Table 1.1 correspond to correlations closer to zero than the classical ones for what concerns the association between altitude and temperatures.

Let us now summarize measurements through a lower number of variables. We could do so, for instance, in order to visualize the data in a two dimensional plot. Table 1.2 gives the percentage of explained variability accounted for by the principal components based on the classical correlation matrix and the two robust counterparts being used.

The first two components obtained with the classical analysis explain a large percentage of variability, but the robust techniques show us that there is a strong sensitivity as they are in disagreement. A strong sensitivity can be simply claimed whenever robust and classical techniques give substantially different results. Robust dimension reduction shows that a non negligible part

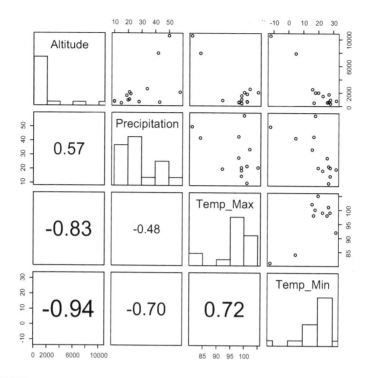

FIGURE 1.3
Butterfly data. Univariate and pairwise distributions of four environmental variables for 16 colonies of *Euphydryas edhita*. Numbers in the lower boxes are correlations.

of the variability accounted for by the first classical component is basically due to outliers: the first component is driven by the outlying colonies.

Figure 1.4 shows the effect of the outliers on the classical analysis. In both panels, the 95% tolerance ellipses have been added. The classical scores (new variables) in the left panel show a linear pattern, with the exception of the two outliers. One outlier falls within the 95% tolerance ellipse (masking) and one observation which does not seem to be unusual falls outside the ellipse (swamping). On the contrary, robust methods lead to a plot where no trend is identified, and the two anomalous points are clearly separated from the rest of the data.

TABLE 1.1
Butterfly data. Robust correlations based on MCD (top) and MM (bottom).

	Altitude	Precipitation	Max. temperature
Precipitation	0.35		
	0.64		
Max. temperature	0.66	-0.21	
	0.45	-0.26	
Min. temperature	-0.62	-0.72	-0.29
	-0.66	-0.91	-0.05

TABLE 1.2
Butterfly data. Variance explained by different principal components methods.

	PC1	PC2	PC3	PC4
Classical	0.79	0.14	0.06	0.01
MCD	0.57	0.34	0.06	0.03
MM	0.62	0.32	0.06	0.00

1.5 An overview of robust sample reduction

There are now many approaches to robust sample reduction. We start focusing on procedures for robust cluster analysis. Most of them are based on the idea of discarding contaminated values. A zero-one weighting scheme is used, where some values are given a zero weight to exclude them from the analysis. This weighting scheme could either be assigned to an entire observation, therefore performing impartial trimming (Gordaliza, 1991; García-Escudero *et al.*, 2010; Neykov *et al.*, 2007), or to part of it, therefore performing snipping (Farcomeni, 2014a,b). In the first case, contaminated observations are entirely removed. They do not contribute to centroid estimation and are not assigned to any cluster. This procedure is usually referred to as *impartial trimming* as we allow any configuration of trimmed values, without specifying any preferential pattern (e.g., that a minimum fraction of values are trimmed within each cluster). In the second case, some observations may not be entirely trimmed, and only few entries are removed from those. Snipping is tailored to tackle the issue of isolated contaminated entries, and allows to classify contaminated subjects based on their clean values. The second part of this book is devoted to review trimming and snipping procedures for spherical and elliptical clusters, either with equal or unequal cluster weights. As said, the basic idea is very simple and is related to discarding some or at least part of some

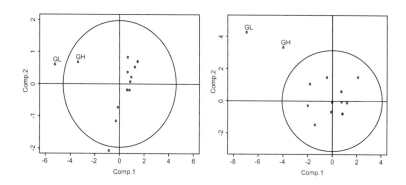

FIGURE 1.4
Butterfly data. First two principal components. Classic (left) and robust (right) based on the MCD.

observations. Refer for instance to Tseng and Wong (2005) for an account of the poor performance that classical clustering algorithms can have given that they forcefully cluster each and every observation.

There are many different other possibilities for sample reduction with unknown labeling, which we only briefly mention here: some approaches are based on augmenting finite mixture models, or on generalizing them to being more flexible, such as Banfield and Raftery (1993) and Frühwirth-Schnatter and Pyne (2010) (and references therein). Other approaches are based on geometric considerations, e.g, Climer and Zhang (2006) and Forero *et al.* (2012).

We now give a brief overview of robust and non-robust clustering procedures on a simple example, based on the data in Figure 1.2. Suppose we would like to divide the Italian regions in two groups based on the two available measurements. R code for this example is not referenced here but it is available from the accompanying website.

The resulting labels as identified by different clustering procedures, which will be introduced and detailed later, are given in Table 1.3. It shall be first of all noted that the group labels obtained with clustering methods are not meaningful *per se*, but are used only to identify regions assigned to the same cluster. For this reason, cluster labels must be read only columnwise to identify regions assigned to the same cluster. Another important annotation regards the presence of zeros in Table 1.3: some robust methods discard some observations entirely. These are not assigned to *any* cluster. This is a very important feature of robust methods, as a cluster is inherently defined by the observations assigned to it. Hence when one or more observations are removed, the interpretation of the results of the analysis overall may be very different.

From Table 1.3 it can be seen that non-robust methods (k-means and Partitioning Around Medoids (PAM)) are rather less stable than robust methods,

TABLE 1.3
Italian demographics data. Groups obtained with different clustering procedures when $k = 2$. A zero indicates that the region is flagged as outlying.

Region	k-means	PAM	tkmeans	tclust	skmeans	sclust
Piemonte	2	1	2	1	1	1
Val d'Aosta	2	1	2	1	1	1
Lombardia	1	1	1	1	2	1
Trentino Alto Adige	1	2	1	2	2	2
Veneto	1	1	1	1	2	1
Friuli	2	1	2	1	1	1
Liguria	2	1	0	0	0	0
Emilia Romagna	2	1	2	1	1	1
Toscana	2	1	2	1	1	1
Umbria	2	1	2	1	1	1
Marche	2	1	2	1	1	1
Lazio	1	1	1	1	2	1
Abruzzo	2	1	2	1	1	1
Molise	2	1	2	1	1	1
Campania	1	2	0	2	2	2
Puglia	1	2	1	2	2	2
Basilicata	1	2	1	2	2	2
Calabria	1	2	1	2	2	2
Sicilia	1	2	1	2	2	2
Sardegna	1	1	1	0	1	1

whose results are slightly more similar. All methods identify more or less one cluster made by the northernmost regions, which are characterized by low birthrate and high mortality, and one made by the southernmost regions, characterized by high birthrate and low mortality. This happens given that all methods identify approximately spherical or at most elliptical clusters. It can be noted that procedures targeting *linear* clusters would here probably lead to a single group made up by all of the regions, with the probable exception of Sardegna (García-Escudero *et al.*, 2009). When we compare the procedures we see that k-means assigns some regions which should belong to the first group (labeled with the number 2 by the algorithm) to the second (which is labeled with number 1 by the algorithm), likely because of the influence of outliers and/or because of the convergence of the algorithm to a sub-optimal solution. PAM does the reverse, by assigning some regions with high birthrate to the first group. Note that PAM labels the first group with a 1 and the second with a 2, but as clarified beforehand, the label itself is silent.

TABLE 1.4

Italian demographics data. Centroids obtained with different clustering procedures when $k = 2$.

Method	Birthrate Cluster 1	Birthrate Cluster 2	Mortality Cluster 1	Mortality Cluster 2
k-means	10.18	7.96	8.83	11.41
PAM	8.22	11.05	10.76	8.62
tkmeans	9.93	8.11	8.91	11.18
tclust	8.31	11.05	10.75	8.62
skmeans	8.12	10.39	10.88	8.90
sclust	8.34	11.05	10.55	8.62

Liguria is identified as being a structural outlier by all of the robust methods, while assigned (correctly) to a cluster composed mostly of regions with low birthrate and high mortality by the non-robust methods. It could be argued as before that Liguria is an extreme but not influential outlier. Both its measurements are extreme, hence it can be classified as a structural outlier. On the other hand, Sardegna is assigned by k-means to the cluster of southern regions because of its low mortality, and by PAM to the cluster of northern regions because of its low birthrate. tkmeans fails to identify Sardegna as an outlier, while tclust correctly flags it. These two methods are based on trimming. The two methods based on snipping (skmeans and sclust) assign Sardegna to the low birthrate group *after* its mortality measurement is flagged (not shown in the table). As a consequence, it can be said that both skmeans and sclust correctly identify Sardegna as being a component-wise outlier. After mortality is removed, there is nothing unusual about Sardegna's birthrate. Note that also the contrary, in this particular case with $p = 2$, is true. The algorithms could have chosen to flag Sardegna's birthrate, but this would have lead apparently to a sub-optimal solution.

Data reduction as said is performed by computing the means of regions assigned to each cluster. In case a robust method is used, the observations flagged as being outlying are not used in this computation. It can be said that cluster centroids for robust methods are often based on the trimmed mean. In Table 1.4 we report the cluster centroid for each of the six methods considered and the two clusters.

There are different uses for the centroids. First of all, we usually have much less previous information about the data, and their most important role is that of guiding the interpretation of the clusters. For instance, from Table 1.4 we can recover our previous interpretation by noting that as far as k-means is concerned, cluster 1 is characterized by a much higher birthrate and by a much lower mortality than cluster 2. In case two centroids were very close with respect to the birthrate, we could say that birthrate was not discriminating (for that particular clustering method). Another possibility is to use the centroid as a new data set with a much smaller sample size. This formally is data reduction and would imply choosing one method (say, `tclust`) and using a new data set with $n = k = 2$ observations, one with birthrate 8.31 and another with birthrate 11.05, and similarly for mortality.

1.6 Example datasets

We conclude this chapter by giving some insights into a few of the example data sets we will use in this book. This section may be seen as a reference for basic data description and motivation when one of the following data sets is used later in the book.

1.6.1 G8 macroeconomic data

A data set regarding the eight most industrialized countries is traditionally used in many graduate and undergraduate level classes. We introduce it here as a very simple but informative case study. We have data regarding $p = 7$ indicators, of slightly different nature, for $n = 8$ countries. The countries involved in the study are France (FRA), Germany (GER), Great Britain (GBR), Italy (ITA), United States of America (USA), Japan (JAP), Canada (CAN); plus Spain (SPA). The variables measured were: Gross Domestic Product index (GDP), Inflation (INF), Budget deficit/GDP (DEF), Public debt/GDP (DEB), Long term interest rate (INT), Trade balance/GDP (TRB), unemployment rate (UNE). The measurements refer to the period 1980-1990, and are of particular interest because most of the variables were considered in the parameters of the Maastricht treaty. European countries are bound to stay within thresholds established in the treaty for some of these variables, otherwise they might be sanctioned by the European Union. The eight countries have slightly different performances with respect to the indicators considered. It may be of interest to summarize the information in order to provide rankings of performance. Furthermore, the countries may be grouped and outliers identified. In particular, it is asked whether Italy can be thought of being an outlier, possibly after standardization. The data are given in Table 1.5.

From Table 1.5 it can be seen that at the time of measurement Spain had lagged behind with respect to GDP, and Italy had a huge deficit and debt. These measurements are likely to be strongly correlated, so that they may be efficiently summarized.

TABLE 1.5
G8 macroeconomic data.

Country	GDP	INF	DEF	DEB	INT	TRB	UNE
FRA	133.40	3.00	−1.50	46.60	10.40	−2.10	8.90
GER	138.80	3.40	−1.90	43.60	6.00	5.90	6.20
GBR	125.10	6.30	−1.30	34.70	11.10	−4.00	6.80
ITA	120.20	7.60	−11.50	100.50	11.90	−0.70	11.20
SPA	92.50	7.30	−3.60	46.80	14.70	−6.50	15.90
USA	176.20	4.30	−2.50	56.20	8.70	−2.70	5.50
JAP	142.00	2.20	2.90	69.80	7.40	1.90	2.10
CAN	166.70	3.30	−4.10	71.90	10.80	1.60	8.10

1.6.2 Handwritten digits data

Optical recognition of handwritten digits and characters is routinely used to solve different problems. A clear example is given by paper mail routing, where address is automatically read on the back of a letter or postcard, which is routed accordingly. Optical Character Recognition (OCR) is the task of translating an image of a text into the text itself. In the previous example, the address read on a letter or postcard is translated into text, which is further translated into a bar code. High definition OCR is used also in banks to deal with checks. OCR might be employed to scan texts for further editing, conservation, and dissemination.

An open issue regards assessing the ability and improving currently available OCR systems. Another important problem regards the evolution of handwriting over time and its variability in different areas, cultures, and languages: if the common handwriting changes, OCR systems need to be updated accordingly.

A data set we will use in this area of research consists in normalized bitmaps of handwritten digits extracted from a preprinted form. A total of 30 subjects contributed by providing $n = 3823$ digits, from 0 to 9. The frequency table of digits is not balanced, with also variability in the number of digits provided by each subject. Data was obtained by scanning the preprinted form and obtaining 32x32 bitmaps. These were divided into non overlapping blocks of 4x4, and the number of pixels were counted in each block. Consequently, each bitmap is translated into 64 measurements, corresponding to a block in the 8x8 input matrix. Each measurement is an integer in the range 0 to 16. It is natural to expect that for all digits some measurements will be quite similar (for instance, we expect low measurements or even zeros in the upper left corner of each bitmap), while other measurements will be widely variable among different digits and therefore will be very useful for correct digit recognition.

Data are freely available from the UCI Machine Learning Repository (Frank and Asuncion, 2010).

The most important issue with these data is in regards to the similarity and dissimilarity among couples of digits. There are 10*9/2=45 possible couples, and we are interested in *directional* errors: a 9 may be easily misclassified as a 0, while the contrary may be less common. Hence there are 45*2=90 OCR tasks we are interested in investigating, to assess what are the most and least common errors, and what shall be the focus to improve the performance of OCR. Another issue regards how to further summarize the information, reducing the 64 measurements to a smaller number, for more efficient storing and quicker processing. Errors in classification will mostly be due to variability *within* bitmaps obtained for the same digit, and outliers which may mislead the training process of OCR systems. These two features are key investigation issues with this data set. As it often happens, we can expect any kind of outliers.

1.6.3   Automobile data

The 1985 Auto Imports Database contains technical and insurance information about $n = 205$ cars collected from the 1985 Automotive Yearbook, Insurance Services Office and Insurance Institute for Highway Safety. There are $p = 26$ variables of which ten are categorical. The data set is publicly available at the UCI Machine Learning Repository (Bache and Lichman, 2013). Data measure length, height and weight of the car, engine characteristics, fuel efficiency, price and some actuarial ratings such as the symboling. Symboling is a proxy of the degree to which the auto is more risky than its price indicates. The symboling process works as follows: cars are initially assigned a risk factor associated with their price. Then, if the crash rate indicates that it is more or less risky than expected, the symbol is adjusted by moving it up (or down) the scale. A value of +3 indicates that the auto is more risky than expected, -3 that it is probably pretty safe. Finally, data measure the relative average loss payment per insured vehicle per year.

Automobile data can be used to identify attributes that explain the differences among the different typologies of cars. This information may be used by insurance policies to set insurance prices. Furthermore, groups of cars are of interest and the identification of outliers may be useful to flag unusual vehicles. Outliers formally coincide with cars deserving specific insurance conditions.

1.6.4   Metallic oxide data

A sampling study was designed to explore the effects of process and measurement variation on properties of metallic oxide lots. The metal content minus 80% by weight was recorded for two types of metallic oxide raw material, in respectively 18 and 13 lots, by two randomly chosen chemists for each sample and two samples from each lot. Data come from Bennet (1954), and were analyzed with a robust mixed model approach by Fellner (1986); Zewotir and

Galpin (2007) and with cluster analysis methods by Farcomeni (2009b) and Farcomeni (2014a).

First of all, these data can be used to discriminate between the two types of metallic oxide material, after controlling for measurement variation due to chemist and lot variables. The main motivating question is whether the two types yield sensibly different metal contents in general. Another issue is in regards to the ability of chemists to precisely measure the metal content: each sample is measured four times, and in principle we should obtain the same measurements on it. Furthermore, what is the expected metal content of a *clean* sample of Type 1, and what the expected metal content of a *clean* sample of Type 2? What is the variability that can be expected? What will be the 5th and 95th percentile of metal content of *clean* samples? A final issue is in regards to the possible presence of outliers: are there lots with unusually low or high metal contents? Do any of the two Types lead to outlying samples more often than the other? The data set is rather small so we report it in Table 1.6.

Visual inspection in this simple example is enough to realize that lots 6 and 7 of Type 2 may be structural outliers, with unusually low metal content in their samples. It may instead be rather more difficult to visually identify component-wise outliers, like for instance the last measurement of lot 2, sample 2, Type 1. The latter may be caused simply to a measurement error by chemist 2 on sample 2. Ignoring these outliers may lead to bias in answering the questions above, in particular regarding expectation and variability of metal content.

1.6.5   Spam detection data

Spam e-mails have various forms, like advertisements of products, services or web sites, pharmaceuticals and stock promotions. Spam often contain also scams, job offers, make-money-fast schemes, chain letters, pornography and so on. These unsolicited e-mails can often be considered as outliers in a collection of regular e-mails coming, for instance, from filed work and personal activities. Spam detection is an open issue because of the constantly pioneering new techniques devised by spammers to bypass e-mail filters. In building an up-to-date spam detection tool, two features need to be taken into account. The first is the ability of the system to detect the actual spam e-mails, the second is its reliability in avoiding the block of legitimate correspondence, that may wrongly be classified as spam. We consider a data set publicly available on the UCI Machine Learning Repository. The variables measure characteristics of the e-mail which could indicate its being a legitimate or spam message. These include counts of occurrences of particular strings, including keywords typical of one or the other group. The last column of the available data set denotes whether the e-mail was considered spam or not by a human reader. The huge collection of data suggest us to apply dimension reduction techniques that are able to compress the available information. The compressed information

TABLE 1.6
Metallic oxide data.

		Sample 1				Sample 2			
Lot	Type	Chemist 1		Chemist 2		Chemist 1		Chemist 2	
1	Type 1	4.1	4.0	4.3	4.3	4.1	4.0	4.1	4.0
2	Type 1	4.1	4.0	4.0	3.9	4.2	4.2	3.7	4.6
3	Type 1	3.5	3.5	3.4	3.6	3.4	3.3	4.0	3.5
4	Type 1	4.2	4.2	4.2	4.3	4.1	3.7	4.1	4.6
5	Type 1	3.7	3.8	3.3	3.3	3.2	3.1	3.1	3.2
6	Type 1	4.0	4.2	3.8	4.2	4.1	4.3	4.2	4.1
7	Type 1	4.0	3.8	3.8	4.0	3.6	3.8	3.9	3.8
8	Type 1	3.8	3.9	4.0	3.9	4.0	4.0	4.2	4.0
9	Type 1	4.2	4.5	4.3	4.1	3.8	3.7	3.8	3.8
10	Type 1	3.6	4.0	4.0	3.7	3.9	4.1	4.2	3.7
11	Type 1	4.6	4.6	4.0	3.4	4.4	4.5	3.9	4.1
12	Type 1	3.3	2.9	3.2	3.9	2.9	3.7	3.3	3.4
13	Type 1	4.5	4.5	4.0	4.2	3.7	4.0	4.0	3.9
14	Type 1	3.8	3.8	3.5	3.6	4.3	4.1	3.8	3.8
15	Type 1	4.2	4.1	3.8	3.8	3.8	3.8	3.9	3.9
16	Type 1	4.2	3.4	3.7	4.1	4.4	4.5	4.0	4.0
17	Type 1	3.3	3.4	3.9	4.0	2.2	2.3	2.4	2.7
18	Type 1	3.6	3.7	3.6	3.5	4.1	4.0	4.4	4.2
1	Type 2	3.4	3.4	3.6	3.5	3.7	3.5	3.1	3.4
2	Type 2	4.2	4.1	4.3	4.2	4.2	4.2	4.3	4.2
3	Type 2	3.5	3.5	4.2	4.5	3.4	3.7	3.9	4.0
4	Type 2	3.4	3.3	3.6	3.1	4.2	4.2	3.3	3.1
5	Type 2	3.2	2.8	3.1	2.7	3.0	3.0	3.2	2.7
6	Type 2	0.2	0.7	0.8	0.7	0.3	0.4	0.2	−1.0
7	Type 2	0.9	0.6	0.3	0.6	1.0	1.1	0.7	1.0
8	Type 2	3.3	3.5	3.5	3.4	3.9	3.7	3.7	3.7
9	Type 2	2.9	2.6	2.8	2.9	3.1	3.1	2.9	2.7
10	Type 2	3.8	3.8	3.9	3.8	3.4	3.6	4.0	3.8
11	Type 2	3.8	3.4	3.6	3.8	3.8	3.6	3.9	4.0
12	Type 2	3.2	2.5	3.0	3.5	4.3	3.7	3.8	3.8
13	Type 2	3.4	3.4	3.3	3.3	3.5	3.5	3.2	3.3

can be used to gain an insight into the important keywords identifying one or another group, and can then be used to discriminate between regular and spam e-mails.

1.6.6 Video surveillance data

The detection of persons and objects in a cluttered background is of paramount importance in the area of video surveillance. Given a sequence of surveillance video frames, the most basic task in video surveillance is to estimate a good model for the stationary background. This task is complicated by the presence of foreground persons and/or objects, that can be considered anomalies, and, generally, occupy only a fraction of the image pixels. The reconstruction of the background, then, is expected to be robust with respect to these outliers and to make us able to identify those activities that eventually stand out from the background.

The frames have thousands or tens of thousands of pixels, and each video fragment contains hundreds or thousands of frames. Each image is stored in a row vector formed by concatenating each row of image pixels. This massive amount of data is a natural candidate for dimensionality reduction, due to the large correlation between frames. Here, we consider a collection of $n = 506$ video frames of resolution 120×160 pixels that has been gathered from a static camera over two days. Then, the 120×160 resolution turns into $p = 19200$ dimensions. The data are available from www.cs.cmu.edu/~ftorre.

1.6.7 Water treatment plant data

Monitoring the state of operating plants is needed in order to prevent, identify, and react to faults. A multidimensional evaluation of the state at each of many locations is often needed in order to monitor the process. An example regards treatment of waste waters. Data have been collected by the UCI Machine Learning Repository and contain $n = 380$ daily measurements for $p = 38$ variables. These include measurements of levels of general flow, Zinc, pH, chemical and biological demand of oxygen, suspended and volatile solids, sediments, and conductivity. These variables are simultaneously measured at input, first and second settler, and output. Also global measures of performance are obtained.

The main objective is to classify the operational state of the plant, identify trends and interdependence among input, settlers and output, identify structural outliers with special attention to moments of high load, and the consequent measurement at each stage of the treatment of a high input flow. Occasional measurement errors should be tackled in order to avoid bias in the analysis, where component-wise outliers may be given by anomalous situations at one or few locations for one or more variables.

2

Multivariate Estimation Methods

This chapter provides the reader with an overview of some estimation methods and key concepts that are needed to design effective robust procedures for estimation and outlier detection. Some arguments are rather technical. We try to present them in a simple form, with the aid of appropriate references, R codes and graphical displays.

First, the framework of robust estimation will be introduced for univariate problems. Then, a series of robust estimators of multivariate location and covariance will be introduced. These will be used extensively throughout the book.

The sensitivity of classical location and covariance estimates will be illustrated from different points of view. A consequence of this sensitivity is that all multivariate techniques based on those estimates might be sensitive as well and lead to unreliable conclusions. Contamination may bias location estimates, covariance estimates may be inflated, tolerance ellipses might be too large and unrealistically oriented or centered. Formally, tolerance ellipses are regions of estimated density. A 95% tolerance ellipse, for instance, is a region of a multivariate space which is expected to contain 95% of the measurements at population level, that is, from any sample selected uniformly at random from the population of interest.

In many multivariate techniques multivariate normality is assumed. A similar and only slightly more general assumption is that the data generating density is elliptically contoured.

A first requirement of robust estimators is efficiency at the specified model distribution, meaning that only a bounded and possibly pre-specified loss of efficiency is expected in absence of contamination. The efficency loss can be measured as the ratio of the variance of the robust and classical procedures. A second requisite of robust estimators is that they have a bounded influence function. The bias of the estimator can not be arbitrarily large. A third requirement is a high breakdown point: robust estimators need to resist a large amount of contamination before they become unrealistic. This should be possibly combined with a relatively small maximum bias.

Robust estimation methods described in this chapter may be used for two objectives: first, to provide an accurate fit based on the clean part of the data; secondly, to detect outliers. Robust estimation and outlier identification are strictly connected tasks.

Robust estimates are defined by a set of *weights* aiming at down-weighting or trimming those points departing from model assumptions. After using robust estimators, aberrant observations can be detected as outliers by looking at their *distance* from the robust fit. Multivariate outlier detection is often a problematic issue to deal with, since it can be difficult or even impossible to identify outliers by using exploratory data analysis tools, especially with high-dimensional data.

2.1 Robust univariate methods

The purpose of this section is to review the necessary background for robust estimation in the univariate setting. The univariate setting is rather simple and useful to introduce basic and important concepts that will be used throughout the book. These methods will be later extended to multivariate problems.

The following review is not exhaustive but intended as a gentle introduction to some technical issues.

Let us consider an univariate sample $x = (x_1, x_2, \ldots, x_n)$ of size n from a location-scale (two parameter) distribution $F \in \mathcal{F}$ with

$$\mathcal{F} = \{F | F = F(x; \mu, \sigma), \mu \in \mathbb{R}, \sigma > 0\}$$

where μ is the location parameter and σ the scale parameter. The most common case is to assume F is Gaussian, where μ is the expected value and σ^2 the population variance. The sensitivity of the sample mean $\bar{x}$ and standard deviation s has been already mentioned. Alternative robust estimates, such as the median and MAD may be helpful for robust univariate inference.

We now investigate further the properties of these estimators.

2.1 Example (Sample mean and median) *Let us compare the sample mean and median, first. The sample mean can be written as $\bar{x} = \hat{\mu}(F_n) = \int x \, dF_n$ and the corresponding population functional is the expected value $\hat{\mu}(F) = \int x \, dF(x) = \mu$. The sample median is $Med(F_n) = F_n^{-1}\left(\frac{1}{2}\right)$ and its population functional is $Med(F) = F^{-1}\left(\frac{1}{2}\right)$.*

The IF for the mean is straightforward to compute. According to Definition 1.2,

$$\hat{\mu}(F_\epsilon) = \int_{\mathcal{X}} x \, dF_\epsilon(x) = (1 - \epsilon)\mu(F) + \epsilon x$$

and then

$$IF(x; \hat{\mu}, F) = \frac{\partial}{\partial \epsilon} \hat{\mu}(F_\epsilon)_{|\epsilon=0} = x - \mu(F) \ .$$

The median functional $Med(F)$ is the solution to

$$\int_{\mathcal{X}} \text{sign}\,(x - Med(F)) \ dF(x) = 1 - 2F(Med(F)) = 0 \ .$$

Under the ϵ-neighborhood model with a point mass contamination at x, the median is the solution to

$$\int_{\mathcal{X}} \text{sign}\,(x - Med(F_\epsilon))\ dF_\epsilon(x) =$$
$$(1-\epsilon)\,[1 - 2F(Med(F_\epsilon))] + \epsilon\,\text{sign}\,(x - Med(F_\epsilon)) = 0\ .$$

By taking the derivative with respect to ϵ and rearranging terms, we obtain the following expression:

$$IF(x; Med, F) = \frac{\partial}{\partial \epsilon} Med(F_\epsilon)_{|\epsilon=0} = \frac{\text{sign}(x - Med(F))}{2f\,(Med(F))}\ ,$$

where f denotes the density function corresponding to the distribution function F. When the model is the standard normal, i.e., $F = \Phi$, then

$$IF(x; \hat{\mu}, \Phi) = x$$
$$IF(x; Med, \Phi) = \sqrt{\frac{\pi}{2}}\text{sign}(x)\ .$$

Consequently, unlike the sample mean, the median has a bounded IF. Furthermore, we could arbitrarily increase (decrease) any value above (below) the median without breaking down: the asymptotic BP of the median is 50%.

Once we have derived the expression for the IF, we can investigate the asymptotic behavior of the estimators and in particular the asymptotic efficiency of the median relative to the sample mean (which corresponds to the maximum likelihood estimator). First, we obtain the expression for the asymptotic variances according to Proposition 1.1: at the standard normal distribution, $ASV(\bar{x}, \Phi) = 1$ and $ASV(Med, \Phi) = \frac{\pi}{2}$. The asymptotic efficiency of the median is therefore measured by the reciprocal of its asymptotic variance.

$$ARE(Med, \Phi) = \frac{1}{ASV(Med, \Phi)} = \frac{2}{\pi} = 0.637\ .$$

The robustness of the median can be investigated further through the evaluation of its maxbias curve (Huber, 1964)

$$MB(Med, F, \epsilon) = F^{-1}\left(\frac{1}{2(1-\epsilon)}\right), 0 \le \epsilon \le 0.5\ .$$

For a detailed account on maxbias curves for robust location estimators see also Croux and Haesbroeck (2002).

2.1.1 M estimators

M estimators are derived in order to have a bounded influence estimator for the location parameter, with a larger asymptotic efficiency than the median.

M estimators will also have a reasonable bias under contamination. Actually, the family of M estimators is a very flexible family that includes the sample mean and median as special cases.

An M estimate $\hat{\mu}_M$ for the location parameter is defined as the solution to the estimating equation

$$\sum_{i=1}^{n} \psi\left(\frac{x_i - \mu}{\sigma}\right) = 0, \tag{2.1}$$

where $\psi(\cdot)$ is a known real valued function. Here we assume that the scale parameter is known and set $\sigma = 1$. The corresponding population M-functional $\mu_M(F)$ is implicitly defined by

$$E_F\left[\psi(X - \mu_M(F))\right] = \int_{\mathcal{X}} \psi(X - \mu_M(F)) \, dF(x) = 0 \,. \tag{2.2}$$

If ψ is the identity function the sample mean is obtained, whereas when ψ is the sign function we obtain the median. When the ψ function satisfies (2.2) at the specified model F, the estimating equation (2.1) is called *unbiased*. This is a fundamental property since unbiased estimating equation imply consistency (Hampel *et al.*, 1986).

The influence function of an M estimator is

$$IF(x; \hat{\mu}_M, F) = M^{-1}\psi(X - \mu) \tag{2.3}$$

where $M = -E_F\left[\psi'(x - \mu)\right]$ and ψ' denotes the first derivative of ψ. Hence, any choice of a bounded ψ function yields an M estimator with a bounded IF, that is, formally locally robust.

According to (1.7), the asymptotic variance of $\hat{\mu}_M$ is

$$ASV(\hat{\mu}_M, F) = \frac{\Omega}{M^2},$$

with $\Omega = E_F\left[\psi(X - \mu)^2\right]$. Its reciprocal gives the $ARE(\hat{\mu}_M, F)$.

The maxbias curve of an M estimator (Martin and Zamar, 1993) with known scale and a continuous, monotone, odd and bounded ψ-function is

$$MB(\hat{\mu}_M, F, \epsilon) = g^{-1}\left[\psi(\infty)\frac{\epsilon}{1 - \epsilon}\right] \,,$$

where $g^{-1}(\cdot)$ denotes the inverse function of $g(t) = E_F\left[\psi(X - t)\right]$.

2.1.2   Huber estimator

A popular choice for ψ is given by the family of Huber functions

$$\psi_H(t) = \begin{cases} t & |t| \leq c \\ c \, \text{sign}(t) & |t| > c \end{cases} \tag{2.4}$$

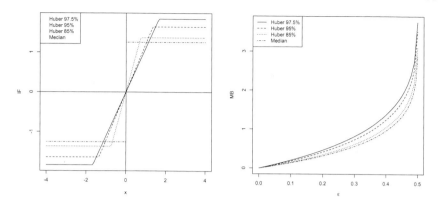

FIGURE 2.1
Huber estimator of location. Influence function (left) and maxbias curve
(right) for different efficiencies.

The parameter c can be used to trade-off robustness and efficiency. In fact,
one can fix a desired efficiency level eff and then fix the constant c by solving
the equation $ARE(\hat{\mu}_H, F) = eff$. It shall also be noted that $c = 0$ leads to
the median, whereas for large values of c the difference between $\hat{\mu}_H$ and $\bar{x}$ is
negligible (the sample mean corresponds to the case $c \to \infty$).

At the normal model, the efficiency of the Huber estimator of location $\hat{\mu}_H$
is

$$ARE(\hat{\mu}_H, \Phi) = \frac{[2\Phi(c) - 1]^2}{2c^2[1 - \Phi(c)] + 2\Phi(c) - 1 - 2c\phi(c)} \cdot \qquad (2.5)$$

The robustness properties of the Huber estimator of location are illustrated in
Figure 2.1: the left panel displays $IF(x; \hat{\mu}_H, \Phi)$ for values of c corresponding
to different efficiency levels, whereas the right panel shows $MB(\hat{\mu}_H, \Phi, \epsilon)$. In
both panels we also include the median for comparison and reference.

The efficiency of $\hat{\mu}_H$ can be explored by looking at Figure 2.2. The left
panel displays the relationship between asymptotic efficiency and gross error sensitivity of $\hat{\mu}_H$ as a function of c. Robustness and efficiency are enemy
brothers: a small sensitivity means a large robustness with respect to anomalous values but also a small efficiency at the model assumed. The right panel
displays $ARE(\hat{\mu}_H, \Phi)$ as a function of c.

2.1.3 Redescending M estimators

Another popular choice for robust univariate estimation is the family of
Tukey's bisquare (or biweight) functions

$$\psi_T(t) = \begin{cases} t\left[1 - \left(\frac{t}{c}\right)^2\right]^2 & |t| \leq c \\ 0 & |t| > c \end{cases} \qquad (2.6)$$

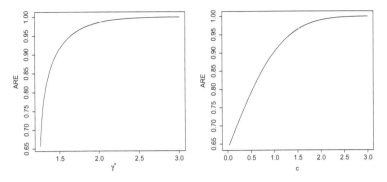

FIGURE 2.2
Huber estimator of location. ARE as a function of the gross error sensitivity
(left) and of the tuning constant c (right).

The main difference with (2.4) is that $\psi_T(t)$ vanishes outside the interval
$[-c, c]$, as illustrated in the left panel of Figure 2.3.

As well as before, by varying c we can attain a certain efficiency at the
assumed model and robustness to the occurrence of outliers. Under normality,
the ARE of the Tukey's estimator of location $\hat{\mu}_T$ is

$$ARE(\hat{\mu}_T, \Phi) = \frac{[2\Phi(c) - 1 - \frac{6}{c^2}h_c(2) + \frac{5}{c^4}h_c(4)]^2}{h_c(2) - \frac{4}{c^2}h_c(4) + \frac{6}{c^4}h_c(6) - \frac{4}{c^6}h_c(8) + \frac{1}{c^8}h_c(10)} \qquad (2.7)$$

where (Riani *et al.*, 2014b)

$$h_c(k) = \int_{-c}^{c} x^k d\Phi(x) = (k-1)!!F_{\chi^2_{k+1}}(c^2).$$

In the formula above, $k!! = \prod_{j=0}^{\lceil k/2 \rceil - 1}(k - 2j)$ is the semifactorial function,
and $F_{\chi^2_k}(t)$ denotes the distribution function of a χ^2 random variable with k
degrees of freedom, computed at t. The $ARE(\hat{\mu}_T, \Phi)$ is shown in Figure 2.4
as a function of the tuning constant c: by increasing c the estimator becomes
more efficient but less robust.

Finally, we notice that both $\hat{\mu}_H$ and $\hat{\mu}_T$ can be written as a *weighted* mean
with data driven weights of the form

$$\frac{\sum_{i=1}^{n} x_i w_i}{\sum_{i=1}^{n} w_i} \qquad (2.8)$$

where $w_i = \frac{\psi(x_i - \mu)}{x_i - \mu}$. Observations more distant from the center receive smaller
weights. The weight functions $w(x) = \frac{\psi(x)}{x}$ corresponding to ψ_H and ψ_T are
shown in the right panel of Figure 2.3.

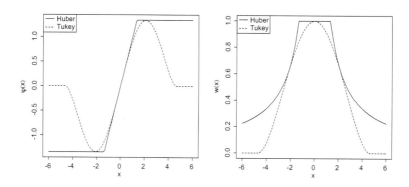

FIGURE 2.3
Huber and Tukey estimators. Left: estimating functions. Right: weight functions. Tuning constants have been chosen to attain 95% efficiency at the normal model.

2.2 R Illustration (Huber and Tukey ψ functions) *The estimating functions plot in the left panel of Figure 2.3 were computed with the following* R *code.*

```
> library(MASS)
> library(robustbase)
> curve(x*psi.huber(x,k=1.345),
+ from=-6,to=6, ylab=expression(psi(x)))
> curve(Mpsi(x,psi='bisquare',cc=4.685),add=T,lty=2)
```

The weight functions plot in the right panel of Figure 2.3 were computed with the following R *code.*

```
> curve(psi.huber(x,k=1.345),from=-6,to=6,
+ ylab=expression(psi(x)),ylim=c(0,1)
> curve(Mwgt(x,psi='bisquare',cc=4.685),add=T,lty=2)
```

The tuning constants k *and* cc *were determined by solving* $ARE(\hat{\mu}_H, \Phi) = 0.95$ *and* $ARE(\hat{\mu}_T, \Phi) = 0.95$, *according to expressions (2.5) and (2.7), respectively. The legend was added to the existing plot by using the function* legend *and specifying the option* 'topleft'.

2.1.4 Scale estimators

M estimators of location, such as those defined by (2.4) and (2.6) have a drawback: they are not *scale invariant*. When the scale parameter σ also

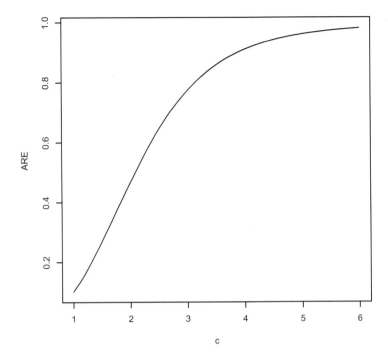

FIGURE 2.4
Tukey estimator of location. *ARE* as a function of the tuning constant c.

needs to be estimated, invariance can be achieved by replacing the unknown parameter by a robust estimate of scale. A robust estimate of location with unknown scale will be the solution to

$$\sum_{i=1}^{n} \psi\left(\frac{x_i - \mu}{\hat{\sigma}}\right) = \sum_{i=1}^{n} \frac{x_i - \mu}{\hat{\sigma}} w\left(\frac{x_i - \mu}{\hat{\sigma}}\right) = 0 . \qquad (2.9)$$

Generally, the use of a preliminary estimate of scale is recommended, rather than simultaneous estimation of location and scale. This is rather counter intuitive, but usually leads to procedures with better robustness properties. The best strategy consists in combining a 50% BP preliminary scale estimate with a high efficient location estimate, even if the ARE of the scale estimator is not sufficiently large. A good solution is the MAD, that in Example 1.1 has been suggested as a robust alternative to the classical standard error. The MAD is defined as

$$\mathrm{MAD} = 1.4826 \ \mathrm{Med}\left(|x_i - \mathrm{Med}(\mathrm{x})|\right) . \qquad (2.10)$$

The multiplicative term $1.4826 = \left[\Phi^{-1}\left(\frac{3}{4}\right)\right]^{-1}$ is obtained imposing consistency to the population standard deviation at the normal model. At the standard normal distribution, the population MAD without the leading constant corresponds to the median of the random variable $|X|$, whose distribution is HalfNormal. This median corresponds to $\Phi^{-1}\left(\frac{3}{4}\right)$. In absence of contamination, it is expected therefore that $MAD(\Phi) \xrightarrow{p} \sigma$. The MAD has 50% BP and its influence function is bounded (Hampel *et al.*, 1986). At the normal model

$$IF(x; MAD, \Phi) = \frac{\text{sign}(|x| - \Phi^{-1}\left(\frac{3}{4}\right))}{4\Phi^{-1}\left(\frac{3}{4}\right)\phi\left(\Phi^{-1}\left(\frac{3}{4}\right)\right)}$$

Nevertheless, its efficiency is low: its ARE at the normal distribution is only 36.74%.

An alternative to MAD is the Q_n (Croux and Rousseeuw, 1992a,b), a robust scale estimator that is more efficient than MAD, still preserving a 50% asymptotic BP. Actually, the ARE of the Q_n estimator is about 82%. This robust estimate of scale is proportional to the .25 quantile of the $\binom{n}{2}$ pairwise differences among couples of data points

$$\{|x_i - x_j|; 1 \leq i < j \geq n\},$$

that is,

$$Q_n = 2.219 q_{0.25}, \tag{2.11}$$

where $q_{0.25}$ is the quantile defined above. The correction term $2.219 = \left[\sqrt{2}\Phi^{-1}\left(\frac{5}{8}\right)\right]^{-1}$ makes the estimator consistent for σ at the normal distribution. It is worth noting additionally that MAD is location based, whereas Q_n is location free. The influence function of the Q_n estimator is

$$IF(x, Q_n, \Phi) = \frac{2.219\left[0.25 - \Phi(x + 2.219^{-1}) + \Phi(x - 2.219^{-1}\right]}{E_\Phi[\phi(X + 2.219^{-1})]}.$$

We point the reader to Rousseeuw and Croux (1993) for details regarding the influence function and maxbias curve of Q_n.

Another family of scale estimators that will be used in the rest of the book is the class of M estimates of scale (that also includes the MAD as a special case). See Maronna *et al.* (2006). Let us assume without loss of generality that $\mu = 0$. An M estimate of scale $\hat{\sigma}_M$ is defined as the solution to

$$\frac{1}{n}\sum_{i=1}^{n}\rho\left(\frac{x_i}{\sigma}\right) = const, \tag{2.12}$$

with $0 < const < \sup_t \rho(t)$. The constant $const$ is fixed so to guarantee consistency at the normal model. In order to obtain this, it is sufficient to set $const$ as the expected value of $\rho(X/\sigma)$ at the normal model.

A popular choice is Tukey's bisquare ρ-function

$$\rho_T(t) = \begin{cases} \frac{c^2}{6}\left\{1 - \left[1 - \left(\frac{t}{c}\right)^2\right]^3\right\} & |t| \leq c \ . \\ \frac{c^2}{6} & |t| > c \end{cases} \tag{2.13}$$

Notice that $\psi_T(t)$ in formula (2.6) is the first derivative of $\rho_T(t)$. The asymptotic BP of scale M estimators is given by

$$\epsilon^* = \min(\alpha, 1 - \alpha), \alpha \in (0, 1),$$

where $\alpha = \frac{const}{\rho_T(c)}$. Hence, $\alpha = 0.5$ implies $\epsilon^* = 0.5$. After fixing the asymptotic BP, one can choose the tuning constant c by requiring consistency at the model, i.e.,

$$\mathrm{E}_F\left[\rho_T\left(\frac{X}{\sigma}\right)\right] = const = \alpha\frac{c^2}{6} \ . \tag{2.14}$$

At the standard normal distribution

$$\mathrm{E}_\Phi\left[\rho_T\left(X\right)\right] = \frac{h_c(2)}{2} - \frac{h_c(4)}{2c^2} + \frac{h_c(6)}{6c^4} + \frac{c^2}{3}[1 - \Phi(c)] \ .$$

The IF of M estimators of scale can be computed according to the general result given by (2.3). The first derivative in the denominator corresponds to

$$-\frac{\partial}{\partial\sigma}\rho\left(\frac{x}{\sigma}\right) = \frac{x}{\sigma^2}\psi\left(\frac{x}{\sigma}\right) \ .$$

Hence, at the standard normal distribution, the influence function is

$$IF(x; \hat{\sigma}_M, \Phi) = \frac{\rho(x) - const}{\mathrm{E}_\Phi[X\psi(X)]} \ .$$

When using the Tukey's bisquare function,

$$\mathrm{E}_\Phi[X\psi_T(X)] = h_2(c) - \frac{2}{c^2}h_c(4) + \frac{1}{c^4}h_c(6) \ .$$

It is worth noting that by varying c, not only the breakdown point but also the efficiency of $\hat{\sigma}_M$ changes: not surprisingly, an inverse relationship between efficiency and BP holds (Riani *et al.*, 2014b).

Figure 2.5 shows the influence functions of MAD, Q_n and M estimates of scale with 50% breakdown point based on $\rho_T(x)$.

2.3 R Illustration (Robust scale estimators) *The influence functions plot in Figure 2.5 were computed with the following R code. The influence function of the Q_n estimator is computed as*

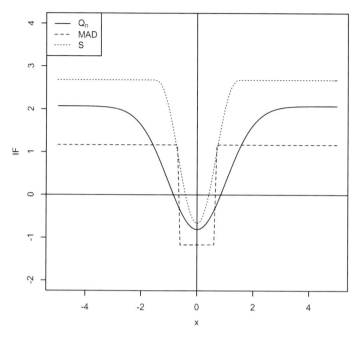

FIGURE 2.5
Influence functions of MAD, Q_n and M estimators of scale.

```
> d<-2.219
> den<-integrate(function(y) dnorm(y+1/d)*dnorm(y),
+ lower=-Inf,upper=Inf)$value
> curve(d*(0.25-pnorm(x+1/d)+pnorm(x-1/d))/den,
+ from=-5,to=5,ylim=c(-2,4),xlab='x',ylab='IF')
```

The influence function of MAD was included as follows:

```
> dmad<-4*qnorm(3/4)*dnorm(qnorm(3/4))
> curve(sign(abs(x)-qnorm(3/4))/dmad,add=T,lty=2)
```

The influence function of the 50% BP S estimator was then included through

```
> alpha<-0.5
> c<-1.547661
> den<-pchisq(c^2,3)-6*pchisq(c^2,5)/(c^2)+
+ 15*pchisq(c^2,6)/(c^4)
> curve(Mchi(x,psi='bisquare',cc=c)-c^2/6*alpha,
+ add=T,lty=3)
```

The tuning constant c was obtained by solving expression (2.14) for $\alpha = 0.5$.

TABLE 2.1
Univariate sample. Huber and Tukey location estimates and 95% confidence intervals for the contaminated sample.

	$\hat{\mu}$	lower	upper
Huber + MAD	25.44	24.24	26.64
Huber + Q_n	25.43	24.24	26.62
Tukey	25.28	24.22	26.34

In all problems in which a monotone ψ function, such as the Huber function ψ_H, is used, $\hat{\sigma}$ in equation (2.9) could be either the MAD or the Q_n estimate. See Rousseeuw and Croux (1992) and Rousseeuw and Croux (1993) for a detailed survey on scale estimators.

On the other hand, M estimates of scale with a large BP are used as preliminary estimates of dispersion when computing the Tukey estimate of location. Hence, first we evaluate $\hat{\sigma}_M$ by (2.12) and select $c = c_0$ for a fixed BP according to (2.14). Then, $\hat{\mu}_T$ is computed as the solution to equation (2.9) with $\psi = \psi_T$, but now using a different tuning constant $c_1 > c_0$ leading to a larger prescribed efficiency than that corresponding to c_0.

2.4 Example (Robust estimation for 1-d Gaussian data) *Let us consider the univariate sample of Example 1.1. The 95% asymptotic confidence interval for μ based on the median and MAD has been obtained as follows*

$$Med(x) \mp z_{.975} \frac{MAD(x)}{\sqrt{n}} \sqrt{ASV(Med, \Phi)} \ .$$

In a similar fashion, we compare the results obtained with the Huber and the Tukey estimator with 95% efficiency at the normal distribution. In the former case we use both the MAD and the Q_n as preliminary scale estimates, in the latter we fix a 50% BP M scale estimate. Actually, the standard errors of $\hat{\mu}_H$ and $\hat{\mu}_T$ are computed by using an estimate of the variance rather than its asymptotic value. The empirical ASV is obtained as $\frac{\hat{\Omega}}{\hat{M}^2} = \frac{ave_i[\psi^2(z_i)]}{ave_i[\psi'(z_i)]^2}$ where $z = \frac{x_i - \hat{\mu}}{\hat{\sigma}}$ and ave denotes the average.

Table 2.1 shows the results for the contaminated sample. The interval based on $\hat{\mu}_T$ is the only one to change slightly from the uncontaminated scenario (24.37, 26.59), whereas those based on $\hat{\mu}_H$ do not change. Notice that all the intervals are narrower than those based on the median because of the gain in efficiency. Huber estimation has been accomplished by calling the function `huberM`, Tukey estimation by using the function `lmrob`, both from the R package `robustbase`.

2.1.5   Measuring outlyingness

Let $(\hat{\mu}, \hat{\sigma})$ be robust estimates of location and scale. They are expected to be informative about the majority of the data and not be influenced by the

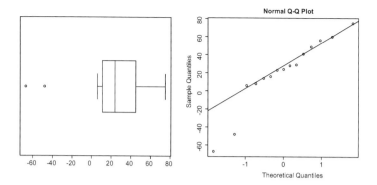

FIGURE 2.6
Darwin data. Boxplot (left) and normal quantile-quantile plot.

presence of anomalous values. Then, outliers can be detected by looking at the univariate measure of outlyingness

$$d_i = \frac{x_i - \hat{\mu}}{\hat{\sigma}} \,. \tag{2.15}$$

Large values of $|d_i|$ will identify outliers. A common practical approach makes use of the standard normal quantiles: a data point can be classified as an outlier if $|d_i| > z_\alpha$.

2.5 Example (Univariate outlier identification) *The following data are differences in heights for 15 pairs of plants. Each pair consisted of one cross-fertilized plant and one self-fertilized plant which germinated at the same time and grew in the same spot. The data were collected by Charles Darwin (Fisher, 1935) and are*

$$
\begin{array}{rrrrr}
49 & -67 & 8 & 16 & 6 \\
23 & 28 & 41 & 14 & 29 \\
56 & 24 & 75 & 60 & -48
\end{array}
$$

The boxplot and the normal quantile-quantile plot in Figure 2.6 indicate that the two negative observations are suspicious. In order to classify them as outliers, we compute the relative distance from the robust location estimate (2.15). Robust estimates are based on the Huber estimator of location with preliminary MAD scale estimation. We compare the results with those arising from a non-robust analysis based on sample mean and standard deviation. From Figure 2.7 we see that in the standard non-robust approach observation -48 is masked for all quantiles $(z_{.975}, z_{.995})$. This is a consequence of the fact that the two negative outliers have attracted the mean and inflated the sample variance of the data. On the contrary, the two outliers are clearly visible when the robust estimates are used.

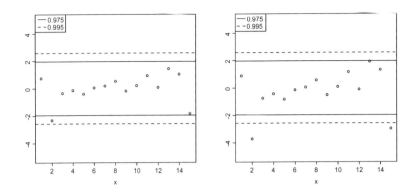

FIGURE 2.7
Darwin data. Classical (left) and robust (right) standardized values.

2.2 Classical multivariate estimation

The sample mean vector and the covariance (correlation) matrix are standard tools for describing location, variability and pairwise dependence in the data. Let $X = (x_1, x_2, \ldots, x_n)^\mathrm{T}$ be a random sample of size n from a multivariate distribution with the ith observation $x_i = (x_{i1}, x_{i2}, \ldots, x_{ip})^\mathrm{T} \in \mathbb{R}^p$, $i = 1, 2, \ldots, n$, $p > 1$. The data are usually stored in an $n \times p$ matrix X, where n stands for the number of observations and p for the number of variables. The $p \times p$ sample covariance matrix is given by

$$S = \frac{1}{n} \sum_{i=1}^{n} (x_i - \bar{x})(x_i - \bar{x})^\mathrm{T},$$

where $\bar{x} = (\bar{x}_1, \bar{x}_2, \ldots, \bar{x}_p)^\mathrm{T}$, with

$$\bar{x}_j = \frac{1}{n} \sum_{i=1}^{n} x_{ij}, j = 1, 2, \ldots, p$$

is the vector of sample means. The primary diffusion of the estimator $(\bar{x}, S)$ in classical multivariate inference arises from its coincidence with the maximum likelihood estimate (MLE) of the parameters of a p−variate Gaussian distribution $N_p(\mu, \Sigma)$. A large majority of multivariate techniques is based on this distributional assumption, even if it may be too restrictive for the sample at hand. Especially for small to moderate sample sizes, the unbiased version of S is preferred, $S_u = \frac{n}{n-1}S$. The estimate $(\bar{x}, S_u)$ can be computed with the basic R functions `colMeans` and `cov`.

In presence of contamination, the use of $(\bar{x}, S)$ can be dangerous and misleading since even one single outlier can badly affect estimation and inference on (μ, Σ) and related quantities, such as eigenvalues and eigenvectors of S. As in the univariate setting, the lack of robustness of $(\bar{x}, S)$ is evident from the unboundedness of their influence functions. Let

$$\bar{x} = \hat{\mu}(F) \quad = \quad \int x \, dF(x)$$

$$S = \hat{\Sigma}(F) \quad = \quad \int (x - \hat{\mu}(F))(x - \hat{\mu}(F))^{\mathrm{T}} \, dF(x).$$

Then, the influence functions at the model F are

$$IF(x; \bar{x}, F) \quad = \quad x - \mu(F)$$
$$IF(x; S, F) \quad = \quad (x - \mu(F))(x - \mu(F))^{\mathrm{T}} - \Sigma,$$

that are clearly unbounded as a function of x: one single outlier can lead to an arbitrary amount of bias. Moreover, it is intuitive that one single outlier can completely destroy the functionals, which means that $(\bar{x}, S)$ have a null asymptotic breakdown point. The deleterious effect of one single outlier on quantities of interest that are functions of $(\bar{x}, S)$ can be investigated by using the *sensitivity curve* (1.3), as in the next example.

2.6 Example (Sensitivity curve for 2-d Gaussian data) *Figure 2.6 displays the sensitivity curves for the Euclidean norm of $\bar{x}$, the determinant of S_u and the ratio of its largest to the smallest eigenvalue λ_1/λ_2, respectively, when one single value is added to the sample from Example 1.2. The added value, x_0, is located at $(x_{01}, 0)$ and x_{01} is allowed to vary in the range $[-10, 10]$. The curves are all unbounded which means that the outlier leads to arbitrary bias for all the functionals we have considered.*

2.3 Robust multivariate estimation

In Example 1.2, it has been shown that the use of a robust estimate of multivariate location and covariance is protective from the deleterious effects of outliers. As in the univariate setting, robust multivariate estimators can be designed to fulfill requirements, such as consistency, efficiency at the assumed sampling model and high breakdown point (BP). In the following, the results on efficiency and breakdown point presented in section 2.1 are extended to multivariate analysis.

An additional reasonable requirement is that of *affine equivariance*. Affine equivariant robust estimators are defined as follows:

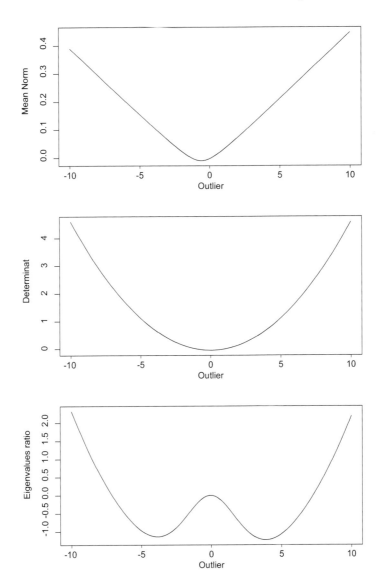

FIGURE 2.8
Bivariate data. Sensitivity curves for $||\bar{x}||$, $|S_u|$ and λ_1/λ_2

Definition 2.1 *A location estimator $\hat{\mu}(X)$ is affine equivariant if and only if (iff) for any vector $b \in \mathbb{R}^p$ and any non singular $p \times p$ matrix A*

$$\hat{\mu}(XA + b) = \hat{\mu}(X)A + b \ .$$

A covariance estimator $\hat{\Sigma}(X)$ is affine equivariant iff for any vector $b \in \mathbb{R}^p$ and any non singular $p \times p$ matrix A

$$\hat{\Sigma}(XA + b) = A^{\mathrm{T}}\hat{\Sigma}(X)A .$$

In common words, inference based on affine equivariant estimators does not change if data are shifted, rotated or if there are changes of measurement scale (e.g., unit of measurement).

In parallel with the univariate scenario, robust multivariate estimates of location and covariance are also defined through a set of weights. These are used to downweight observations lying far from the majority of the data. Weights depend upon the Mahalanobis distance.

Definition 2.2 *The Mahalanobis distance between the the $p-$dimensional data point $x = (x_1, x_2, \ldots, x_p)^{\mathrm{T}}$ and the location vector $\mu = (\mu_1, \mu_2, \ldots, \mu_p)^{\mathrm{T}}$ with respect to the covariance matrix Σ is*

$$d(x; \mu, \Sigma) = \sqrt{(x - \mu)^{\mathrm{T}}\Sigma^{-1}(x - \mu)} . \qquad (2.16)$$

2.3.1 Multivariate M estimators

Under the assumption that the data arise from a $N_p(\mu, \Sigma)$ distribution, where now $\mu = (\mu_1, \mu_2, \ldots, \mu_p)^{\mathrm{T}}$ and Σ is a $p \times p$ symmetric and positive definite matrix, the MLE $(\bar{x}, S)$ is the solution to the likelihood equations

$$\left\{ \begin{array}{l} \sum_{i=1}^{n}(x_i - \mu) = 0 \\ \sum_{i=1}^{n}\left[(x_i - \bar{x})(x_i - \bar{x})^T - \Sigma\right] = 0 . \end{array} \right. \qquad (2.17)$$

The problem of finding a robust estimate for (μ, Σ) can be addressed by replacing the usual score functions with different ψ functions leading to bounded influence estimators.

A first solution is given by the class of *multivariate* M estimators (Hampel *et al.*, 1986). In a fashion similar to the univariate setting, the M estimates $(\hat{\mu}_M, \hat{\Sigma}_M)$ are a *weighted* mean vector and covariance matrix, respectively. Formally, a multivariate M estimate of location and covariance $(\hat{\mu}_M, \hat{\Sigma}_M)$, now with $\hat{\mu}_M = (\hat{\mu}_{M;1}, \hat{\mu}_{M;2}, \ldots, \hat{\mu}_{M;p})^{\mathrm{T}}$,(Maronna, 1976; Huber, 1981) is defined as the solution to

$$\sum_{i=1}^{n}\psi_\mu(x_i; \mu, \Sigma) = \sum_{i=1}^{n}(x_i - \mu)w_{i1} = 0 \qquad (2.18)$$

$$\sum_{i=1}^{n}\psi_\Sigma(x_i; \mu, \Sigma) = \sum_{i=1}^{n}\left[(x_i - \mu)(x_i - \mu)^T w_{i2} - \Sigma\right] = 0 ,$$

where the two weight vectors w_{i1} and w_{i2} depend upon $d_i = d(x_i; \hat{\mu}_M, \hat{\Sigma}_M)$. The weight functions are not necessarily equal: for instance they could correspond to different choices for c in ψ_H. The weights are designed to be small for those observations whose robust Mahalanobis distance is too large compared to the majority of the data. The Mahalanobis distance from a robust fit is commonly referred to as the *robust distance*. Robust distances are the most important tool for multivariate outlier detection, as will be illustrated in Section 2.4.

The robustness properties and the efficiency of the location vector estimator $\hat{\mu}_M$ and of the covariance estimator $\hat{\Sigma}_M$ can be investigated by means of their influence functions. In particular, the influence function of the location M estimator (Hampel *et al.*, 1986) is

$$IF(x; \hat{\mu}_M, F) = \frac{1}{pM} \psi_\mu(X; 0, I_p) \qquad (2.19)$$

with $M = \mathrm{E}_F \left[-\frac{\partial}{\partial \mu^{\mathrm{T}}} \psi_\mu(x_i; \mu, \Sigma) \right] = \mathrm{E}_F \left[w_i(p-1) + \psi'_\mu(x_i; \mu, \Sigma) \right]$. When $p = 1$ we get back the univariate influence function of M estimators.

When using ψ_H or ψ_T we find the tuning constants for a given efficency. The main difference with the univariate framework is that ARE now also depends on the dimension p.

The computation of the IF for the covariance component depends on the following lemma established in Croux and Haesbroeck (2000) (see also Huber and Ronchetti (2009) for further details).

Lemma 2.1 *The influence function of any affine equivariant covariance matrix functional $\Sigma(F)$ is characterized by two functions α_Σ, β_Σ such that*

$$IF(x_i; \Sigma, F) = \alpha_\Sigma(d_i)(x_i - \mu)(x_i - \mu)^{\mathrm{T}} - \beta_\Sigma(d_i)\Sigma$$

The quantities that are needed for the IF of the M estimator of covariance have been derived in Huber (1981).

Some care is needed in evaluating the BP of multivariate M estimators, especially for what concerns the covariance component. Yohai and Maronna (1990) defined the asymptotic bias for covariance matrices by using the *condition number*, i.e., the ratio of the largest to the smallest eigenvalue, as follows

$$MB(\hat{\Sigma}_M, F, \epsilon) = \sup_{F_\epsilon \in \mathcal{P}(F, \epsilon)} \left\{ \frac{\lambda_1(\hat{\Sigma}_M(F_\epsilon))}{\lambda_p(\hat{\Sigma}_M(F_\epsilon))} \right\} .$$

Then the asymptotic BP is

$$\varepsilon^* = \max \left\{ \epsilon : MB(\hat{\Sigma}_M, F, \epsilon) < \infty \right\}$$

and measures the amount of contamination needed to make the functional *explode* or *implode*. The explosion is related to outliers whose effect is that of making the largest eigenvalue go to infinity, the implosion is due to inliers that shrink the smallest eigenvalue towards zero. It can be shown that (Maronna, 1976; Adrover, 1998) the BP of the covariance part is

$$\varepsilon^* \leq \frac{1}{p+1}.$$

Therefore, the maximal BP may be very low when p is large.

2.3.2 Multivariate S estimators

S estimators (Davies, 1987; Lopuhaä, 1989) are defined as the solution $(\hat{\mu}_S, \hat{\Sigma}_S)$ to the problem of minimizing the determinant of $|\hat{\Sigma}|$ subject to the bound

$$\frac{1}{n} \sum_{i=1}^{n} \rho(d_i) = const, \ 0 < const < \sup_{d} \rho(d) \tag{2.20}$$

where $\rho(d)$ is an opportunely specified function: a popular choice is Tukey's bisquare function (2.13). Once again, among the possible choices for *const* in practice only a single function is used, the one guaranteeing consistency at the normal model. This will correspond, as before, to the expected value of $\rho(d_i)$ at the normal model.

The tuning constant c can be chosen as a function of the BP once the BP has been fixed, through (2.14). At the multivariate standard normal distribution $\Phi_p = N(0, I_p)$ (Riani *et al.*, 2014b),

$$\mathrm{E}_{\Phi_p}[\rho_T(X)] = \frac{H_c(2)}{2} - \frac{H_c(4)}{2c^2} + \frac{H_c(6)}{6c^4} + \frac{c^2}{6p}[1 - F_{\chi_p^2}(c^2)] , \tag{2.21}$$

where

$$H_c(k) = \prod_{j=0}^{\frac{k}{2}-1} (p+2j)F_{\chi_{p+k}^2}(c^2).$$

Note that when $p = 1$ the formula above corresponds to the expression seen in the univariate case.

S estimators can be also defined as the solution of M-type estimating equations and their robustness and efficiency properties can be assessed according to (2.19) and Lemma 2.1 (Lopuhaä, 1989; Croux and Haesbroeck, 1999, 2000).

The tuning constant c determines not only the breakdown point but also the asymptotic efficiency of the location and covariance estimators (Riani *et al.*, 2014b). Therefore, some care is needed once we have decided to fix c according to efficiency or breakdown arguments. That is to say, if we choose c to achieve a prescribed efficiency we need to check the corresponding BP, and vice versa.

Figure 2.9 shows the inverse relationship between the ARE of the location vector (whose expression can be found in Riani *et al.* (2014b)) and the BP of the multivariate S estimator. We also note that efficiency becomes independent of the BP as p is increased.

A recent application of S estimation techniques can be found in Farcomeni and Greco (2014). S estimators can be computed in R with the `CovSest` function from library `rrcov`. The desired breakdown point can be specified using the appropriate option.

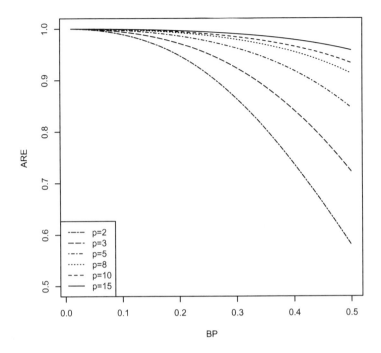

FIGURE 2.9
S estimators. Efficiency of the location estimator as a function of the break-down point and the number of variables.

2.3.3   Multivariate MM estimators

Multivariate MM estimators of location and covariance (Tatsuoka and Tyler, 2000) have been conceived to achieve both high BP and high efficiency at the specified model. They improve upon an initial S estimate of the covariance matrix $\hat{\Sigma}_0$. Let $\hat{\sigma}_0 = |\hat{\Sigma}_0|^{\frac{1}{2p}}$.

A multivariate MM estimator of location and shape $(\hat{\mu}_{MM}, \hat{\Gamma}_{MM})$ (Salibian-Barrera *et al.*, 2006) minimizes

$$\frac{1}{n}\sum_{i=1}^{n}\rho_1\left(d_i\right) = const, \qquad (2.22)$$

where $d_i = \left[(y_i - \mu)\Sigma^{-1}(y_i - \mu)\right]^{1/2}$ and $\Sigma = \hat{\sigma}_0^2\Gamma$, under the constraint $|\Gamma| = 1$. In words, a robust scale estimate is held fixed, while we minimize the objective function with respect to multivariate estimates of location and shape.

TABLE 2.2
MM estimators. Tuning constants for fixed BP and location efficiency for different dimensions.

		p					
		2	3	5	8	10	15
	BP						
c_0	50%	2.66	3.45	4.65	6.02	6.78	8.38
	25%	4.43	5.53	7.24	9.23	10.35	12.72
	ARE						
c_1	85%	3.83	4.15	4.68	5.32	5.68	6.45
	90%	4.28	4.62	5.17	5.84	6.21	7.00
	95%	5.12	5.49	6.10	6.82	7.22	8.08

The corresponding MM estimate for Σ is $\hat{\Sigma}_{MM} = \hat{\sigma}_0^2 \hat{\Gamma}_{MM}$. The S estimate is defined by a $\rho_{T;0}$ whose tuning constant c_0 is set to attain a prescribed BP (generally 50%). Then, the MM estimate is evaluated by using a second bisquare function $\rho_{T;1}$ characterized by a larger tuning constant $c_1 > c_0$, to which a larger efficiency is associated. MM estimators inherit the BP of the initial S estimator but they achieve a larger efficiency. The entries in Table 2.2 give the values of c_0 and c_1 for different dimensions, breakdown points and location efficiencies.

Since the S estimator turns out to be highly efficient in large dimensions, the second step characterizing MM estimation leads improvement in terms of efficiency only in relatively small dimensions ($p \leq 15$). The R function CovMMest from library rrcov computes MM estimates and also returns the initial S estimate: the breakdown point of the initial S estimate and the efficiency of the second step can be tuned using the appropriate options. In particular, efficiency can be specified with regard to the location or the covariance component: in the latter case we refer to it as *shape efficiency* (Salibian-Barrera *et al.*, 2006).

2.3.4   Minimum Covariance Determinant

The Minimum Covariance Determinant (MCD) method introduced by Rousseeuw (1985) estimates (μ, Σ) based on a subset of $n(1 - \varepsilon)$ data points. These data points are chosen as the set whose covariance matrix has the lowest determinant. Informally, they can be deemed as the $n(1 - \varepsilon)$ data points most close to each other, and therefore less likely to be outlying. The final estimate of the mean will simply correspond to the mean of the subset, while the final estimate of the covariance matrix will be proportional to the covariance matrix of the subset. The proportionality constant is chosen to guarantee consistency of the estimator. The method is very popular due to good asymptotic properties, for both the location (Butler *et al.*, 1993) and covariance

(Croux and Haesbroeck, 1999; Cator and Lopuhaä, 2012) component, and due to the availability of fast and efficient algorithms, such as the FASTMCD (Rousseeuw and Van Driessen, 1999).

The main ingredient of the FASTMCD algorithm is the so called C-step (*Concentration* step) which works as follows:

1. Let $\hat{\theta}_0 = (\hat{\mu}_0, \hat{\Sigma}_0)$ be an initial estimate based on a subset of size $n(1 - \varepsilon)$.

2. Calculate the robust distances $d_{0i} = d(x_i, \hat{\theta}_0)$, $i = 1, 2, \ldots, n$.

3. Sort the distances in non increasing order and take a new subset of size $n(1 - \varepsilon)$ based on the lowest distances.

4. Estimate $\hat{\theta}_1 = (\hat{\mu}_1, \hat{\Sigma}_1)$ based on this new subset

At each iteration of the C-step $|\hat{\Sigma}_1| \leq |\hat{\Sigma}_0|$ since the new covariance estimate is based on a subset of the original data characterized by low Mahalanobis distances to the current center. Therefore, by repeating the C-steps iteratively, the sequence of determinants is expected to converge. Convergence will occur in a finite number of steps as there is only a finite number of possible subsets of the desired size. Rousseeuw and Van Driessen (1999) suggest that the initial subset is chosen among several (usually 500) randomly sampled subsets of size $(p+1)$, since such small sized subsets are unlikely to contain outliers. Typical trimming levels are $\varepsilon = .5$ or $\varepsilon = .25$.

Formally, let z be a binary vector such that $\sum z_i = n(1 - \varepsilon)$, where a zero indicates a trimmed observation. The MCD estimator of location and covariance is

$$\hat{\mu}_{MCD} = \frac{1}{\sum_i z_i} \sum_i z_i x_i \tag{2.23}$$

$$\hat{\Sigma}_{MCD} = \frac{c(p, \varepsilon)}{\sum_i z_i - 1} \sum_i z_i (x_i - \hat{\mu}_{MCD})(x_i - \hat{\mu}_{MCD})^{\mathrm{T}},$$

where z is such that

$$|\hat{\Sigma}_{MCD}(z)| \leq |\hat{\Sigma}_{MCD}(z')|, \quad \forall\, z'.$$

The factor

$$c(p, \varepsilon) = \frac{1 - \varepsilon}{F_{\chi^2_{p+2}}(q_{p,1-\varepsilon})} \tag{2.24}$$

makes the MCD consistent at the Normal model by inflating the covariance estimate based on the selected subset. In (2.24), $q_{p,1-\varepsilon}$ denotes the $1 - \varepsilon$-level quantile of a χ^2_p distribution. It shall be additionally noted that a small sample correction factor may reduce the mean squared error of $\hat{\Sigma}_{MCD}$. This is obtained in Pison *et al.* (2002) through numerical approximation methods, and goes beyond the scope of this section. We only mention that in our implementations we have always used the small sample correction factor.

The breakdown properties of the MCD estimator depend on the rate of trimming. It can be shown that MCD has an asymptotic BP $\varepsilon^* = \varepsilon$. When $\varepsilon = 50\%$ this becomes the largest possible (asymptotic) breakdown point for affine equivariant estimators (Lopuhaä and Rousseeuw, 1991; Agulló *et al.*, 2008). The local robustness properties of the MCD are discussed in Croux and Haesbroeck (1999). The R function CovMcd from library rrcov can be used to obtain the MCD.

2.3.5   Reweighted MCD

MCD estimation may be followed by a reweighting step. This leads to a new estimator, the reweighted MCD (RMCD). The reweighting step is performed to increase the finite sample efficiency of the MCD, while keeping a high breakdown point. See Rousseeuw and Leroy (1987).

The reweighting scheme works as follows:

1. compute the robust distances $d_{i;\mathrm{raw}} = d(x_i; \hat{\mu}_{MCD}, \hat{\Sigma}_{MCD})$ based on the *raw* MCD estimates (2.23);

2. Set $z_i = 1$ if $d_{i,\mathrm{raw}}$ is below a threshold (to be defined below), $z_i = 0$ otherwise. The threshold is a function of a decreased trimming level ε'.

3. Update the estimates as

$$\hat{\mu}_{RMCD} = \frac{1}{\sum z_i} \sum_{i=1}^{n} x_i z_i \tag{2.25}$$

$$\hat{\Sigma}_{RMCD} = \frac{c(p, \varepsilon')}{\sum_i z_i - 1} \sum_{i=1}^{n} (x_i - \hat{\mu}_{RMCD})(x_i - \hat{\mu}_{RMCD})^{\mathrm{T}} z_i . \tag{2.26}$$

The basic idea for the RMCD is that a high trimming level when computing the MCD guarantees a high breakdown point. If this was set too high (e.g., there is a proportion of structural outliers substantially lower then ε), at the reweighting step many originally trimmed observations may be included in the estimation set. In this case, reweighting leads to increased efficiency. A common choice for the trimming level at the second step is $\varepsilon' = 0.025$.

The threshold for $d_{i,\mathrm{raw}}$ is usually set according to the $1 - \varepsilon'$ quantile of χ^2_p distribution, which approximates the distribution of the squared Mahalanobis distance.

A slightly better approximation is given by a scaled F distribution as outlined in Hardin and Rocke (2004):

$$d^2(x; \hat{\mu}_{MCD}, \hat{\Sigma}_{MCD}) \sim \frac{pm}{(m - p - 1)} F_{p, m-p+1} , \tag{2.27}$$

where m denotes the unknown degrees of freedom such that $m\hat{\Sigma}_{MCD} \sim$ Wishart$_\mathrm{p}(m, \Sigma)$ holds approximately. In order to fix m, an asymptotic expression m_{asy} has been derived in Croux and Haesbroeck (1999). This expression is based on the IF for the MCD covariance functional. This asymptotic value m_{asy} can be adjusted by the interpolating formula (Hardin and Rocke, 2005)

$$m_{pred} = m_{asy} \exp(0.725 - 0.00663p - 0.078 \log n).$$

This reweighting scheme can be implemented in R by using the output of CovMcd. We provide the resulting CovMcdF function within the online accompanying supplementary material of the book.

After multivariate location and covariance have been estimated by the RMCD described above, the distributional properties of RMCD have been studied in Cerioli (2010). It is shown that a very accurate approximation to the distribution of the squared robust distances is

$$d^2(x_i; \hat{\mu}_{RMCD}, \hat{\Sigma}_{RMCD}) \sim \frac{(\sum z_i - 1)^2}{\sum z_i} \text{Beta}\left(\frac{p}{2}, \frac{\sum z_i - p - 1}{2}\right), \quad (2.28)$$

when $z_i = 1$ and

$$d^2(x_i; \hat{\mu}_{RMCD}, \hat{\Sigma}_{RMCD}) \sim \frac{\sum z_i + 1}{\sum z_i} \frac{(\sum z_i - 1)p}{\sum z_i - p} F_{p, \sum z_i - p} \quad (2.29)$$

otherwise. According to this result, distances have a different distribution depending on the data point is finally trimmed or not. The results above can be used for outlier identification, as will be shown below.

Even if the approximations above provide a substantial improvement, it shall be underlined that the use of the χ_p^2 approximation is still very popular, and characterizes all the R functions that we will use throughout the book. The R function CovMcd from library rrcov can be used to obtain the RMCD based on the most common distributional approximation. More insights about the RMCD will be given in Section 2.4.

2.7 Example (S, MM, RMCD on 2-d Gaussian data) *Let us consider the data used in Example 1.2 and 2.6. Here we aim at comparing the results from S, MM and RMCD estimation and highlight their robustness with respect to the occurrence of outliers. We made use of the default options: the S estimator has 50% BP, the MM estimator is computed with 95% shape efficiency, the MCD has also 50% BP. As well as in Example 2.6, we plot the sensitivity curves for $||\hat{\mu}||$, $|\hat{\Sigma}|$ and λ_1/λ_2 from each method. Even if the SC only provides an approximation to the IF, it is a useful practical tool to study the local robustness of an estimator in all cases in which the computation of the IF may not be simple.*

The SC are plot in Figure 2.7. They are all bounded, meaning that one outlier may only produces small changes in the estimates and the effect of aberrant very large anomalous values vanish, unlike what happened in Figure 2.6. In the middle panel there is only one SC corresponding to S and MM estimators, since their determinants coincide.

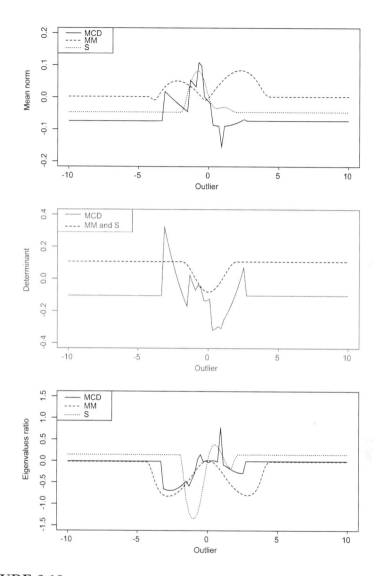

FIGURE 2.10
Bivariate data. Sensitivity curves for $||\hat{\mu}||$, $|\hat{\Sigma}|$ and λ_1/λ_2 from S, MM and MCD estimation.

2.3.6 Other multivariate estimators

The robust multivariate estimators outlined in the previous sections belong to a wide set of robust methods. Some other estimators of multivariate location and covariance have been introduced in the literature and are briefly presented in the following.

The Minimum Volume Ellipsoid (MVE) (Rousseeuw (1985), Davies (1992), Van Aelst and Rousseeuw (2009)) estimator stems from the smallest volume ellipsoid that is based on $n(1-\varepsilon)$ observations. The ellipsoid is centered on the location estimate and its shape is based on the covariance estimate, evaluated on the trimmed set. The MVE is closely connected to the MCD (which is based on the determinant rather than the volume), but the availability of fast algorithms for computing the MCD and its better asymptotic properties have made the latter more popular.

The Stahel-Donoho (SD) estimator (Stahel, 1981; Donoho, 1982; Maronna and Yohai, 1995) was one of the first estimators proposed which could achieve a high BP. The estimator $(\hat{\mu}_{SD}, \hat{\Sigma}_{SD})$ is based on weighted sample mean and covariance matrix. The weights depend on an outlyingness measure based on univariate projections of the data. The fundamental difference with all other procedures seen so far is that univariate projections replace the role of robust distances (Huber, 1985). A univariate projection of an observation x_i along the direction of a given $p \times 1$ vector a is simply its cross-product $x_i^{\mathrm{T}} a$. Let $\hat{\mu}$ and $\hat{\sigma}$ be univariate robust estimates of location and scale, respectively. The outlyingness of x_i can be measured by

$$o(x_i) = \max_{a \in \mathbb{R}^p} \frac{x_i^{\mathrm{T}} a - \hat{\mu}(x_1^{\mathrm{T}} a, x_2^{\mathrm{T}} a, \ldots, x_n^{\mathrm{T}} a)}{\hat{\sigma}(x_1^{\mathrm{T}} a, x_2^{\mathrm{T}} a, \ldots, x_n^{\mathrm{T}} a)} . \tag{2.30}$$

The properties of the SD estimator have been studied in Gervini (2002), Zuo *et al.* (2004) and Debruyne and Hubert (2009). A recent adaptive modification has been given in Van Aelst *et al.* (2011) and one based on Huberized outlyingness in Van Aelst *et al.* (2012). Another well known alternative is the GK estimator (Gnanadesikan and Kettenring, 1972) that is based on *pairwise* robust covariances. Since the resulting covariance estimate may not be positively definite, Maronna and Zamar (2002) proposed an orthogonal version (OGK) which does not suffer from this limitation. The functions for computing these estimates are CovSde and CovOGK, respectively, also available from the R library rrcov.

There are other methods which can be used in multivariate estimation that are based on hard rejection rules, such as impartial trimming (García-Escudero *et al.*, 2008), trimmed maximum likelihood (Cuesta-Albertos *et al.*, 2008) and the Forward Search (Riani *et al.*, 2009, 2012).

A different approach to robust estimation is based on the weighted likelihood approach of Markatou *et al.* (1998). Robust estimates of location, scale and covariance are weighted estimates. The final weights are a smooth function of the Pearson residuals (Lindsay, 1994; Park *et al.*, 2002; Agostinelli and Greco, 2013), that give a measure of the discrepancy between the assumed model and the data.

Finally, all estimators presented above are mostly focused on structural outliers. There are some very recent estimators that can deal with component-wise outliers, including 2GSE (Agostinelli *et al.*, 2014) and `snipEM` (Farcomeni, 2014a, 2015). The latter is a special case of `sclust` (that is, it corresponds to a robust clustering method when $k = 1$). A method that can deal simultaneously with structural and component-wise outliers is `stEM`, which combines trimming and snipping ideas (see the appendix).

2.4 Identification of multivariate outliers

So far we have been mainly concerned with the problem of robust estimation of multivariate location and covariance. A strongly related problem is that of the identification of outliers and of unusual patterns in the data. The basis for multivariate outlier detection is the Mahalanobis distance of Definition 2.2. Outliers can be generally defined as values exceeding a threshold of the Mahalanobis distance.

The classical (non-robust) approach to outlier detection is based on the Mahalanobis distances evaluated at the MLE. Multivariate outliers can, in principle, be identified as those values whose distance $d(x; \bar{x}, S)$ is larger than a fixed threshold. This cut-off value is based on the distribution of the squared Mahalanobis distance under the p-variate Normal model with known location vector and covariance matrix, that is

$$d^2(X; \mu, \Sigma) \sim \chi_p^2 .$$

Nevertheless, the χ_p^2 distributional result holds only approximately when the MLE is plugged in. Therefore, it is suggested to use an exact distributional result $(\bar{x}, S)$ (Gnanadesikan and Kettenring, 1972), stating that

$$d^2(x; \bar{x}, S) \sim \frac{(n-1)^2}{n} Beta\left(\frac{p}{2}, \frac{n-p-1}{2}\right) . \tag{2.31}$$

Hence, cut-off values to detect outliers should be based on the quantiles of a scaled Beta distribution. A very typical choice consists in the use of the .975-level quantile of the selected reference distribution. This approach is flawed in many respects, as $(\bar{x}, S)$ depends on the outliers. The sample mean and sample covariance matrix have been influenced and biased by the outliers themselves. Therefore, if contamination is suspected, the distributional results regarding the Mahalanobis distances at the MLE do not hold any more.

If we identify outliers based on the Mahalanobis distance of observations from the sample mean vector $\bar{x}$ with respect to the sample covariance matrix S, we may have *swamping* and *masking*. These are general problems with any outlier detection procedure. There is masking whenever a false negative occurs, that is, an outlier is not flagged. There is swamping when a false positive occurs, that is, a genuine measurement is flagged as outlying. Of course, no outlier detection procedure can be perfect and a good idea is to try and tune the degree of masking and swamping (Cerioli and Farcomeni, 2011). When we use the MLE, masking and swamping are out of control. Outliers may have small distances due to the attraction of $\bar{x}$ towards them and variance inflation. Therefore they might not be thresholded with high probability (*masking*). On the other hand, genuine data points that are far from the outlying observations may have large distances and be erroneously flagged as outliers (*swamping*). More details on the important concepts of masking and swamping can be found in Davies and Gather (1993). Another problem that may arise with the MLE concerns leverage points: leverage points may be outlying only in one or very few dimensions, therefore having a small Mahalanobis distance overall. Leverage points may violate the general association pattern of the data without even being outlying in any direction.

Many of the problems listed above can be avoided by computing robust Mahalanobis distances (Filzmoser *et al.*, 2008; Cerioli, 2010). Actually, robust estimates of location and covariance are not attracted towards the outliers, which, therefore, will have large distances. The only difficult issue concerns evaluation of distributional results for the robust distances. As a matter of fact when a robust estimator replaces the MLE the distribution of the Mahalanobis distances may be different. The most common and simple rule still assumes that standard asymptotic results hold for the robust distances. Therefore an observation will be declared an outlier when its squared robust distance is larger than the $\chi^2_{p;0.975}$ quantile (Rousseeuw and Van Zomeren, 1990). Even if this is the usual practice in applied robust statistics, there is empirical evidence that the use of the χ^2_p may be unsatisfactory in finite samples for S and MM estimators and for the MCD (Cerioli *et al.*, 2013b). A better approximation to the distribution of the Mahalanobis distances for the MCD and RMCD has been reported in the previous section.

2.4.1 Multiple testing strategy

The use of thresholds that are independent of n to assess the outlyingness of a data point based on its robust distance can lead to a large amount of swamping. The rate of *false discoveries* increases because of the *multiplicity* of the detection procedure. Actually, each of the n data points is tested for outlyingness and even if a false rejection might be obtained with very small (say, 2.5%) probability for a specific unit, the probability of having at least one false rejection may easily be bound by $\min(0.025n, 1)$ using Bonferroni inequality.

Formally, the outlier detection problem can be stated in terms of testing n null hypotheses of the form

$$H_{0i} : x_i \sim N_p(\mu, \Sigma) \ .$$

Let $(\hat{\mu}, \hat{\Sigma})$ be a robust estimate of multivariate location and covariance. The test statistic for H_{0i} is based on the robust Mahalanobis distance of x_i, $i = 1, \ldots, n$. Large distances indicate evidence against H_{0i}. The level of the test should take into account the fact that n tests are performed simultaneously. This brings about a multiple testing issue. For a general review of multiple testing see Farcomeni (2008). A careful control of the level (probability of Type I error) of each test leads to a control of the expected swamping. This is in general not crucial, but very important in certain applications where outlier identification is connected with costly and sometimes risky procedures or consequences. An overview of some applications that focus on accurate outlier detection is given in Cerioli and Farcomeni (2011), where the general idea of this section is discussed in depth. In Cerioli and Farcomeni (2011), formal procedures for detecting multivariate outliers are introduced based on $d(x_i; \hat{\mu}_{RMCD}, \hat{\Sigma}_{RMCD})$. These can be directly extended to other multivariate robust estimates.

Formal control of swamping can in summary be accomplished by employing adjustments commonly used in the different context of multiple testing. Effective rules may involve Bonferroni correction, that is, computing $(2.5/n)\%$ rather than 2.5% thresholds. Bonferroni correction guarantees that no false detection will occurr with high probability (e.g., 97.5% overall). Different rules can be based on the False Discovery Rate (FDR) of Benjamini and Hochberg (1995) or the False Discovery eXceedance (FDX) of Lehmann and Romano (2004) and van der Laan *et al.* (2004). This will guarantee a small proportion of false detections among the detected outliers. These procedures are more appropriate for large sample size and/or when the consequences of a false detection are not dramatic. In order to control the FDR or FDX, the distributional results about $d^2(x_i; \hat{\mu}, \hat{\Sigma})$ should be used to compute a p-value, that is,

$$p_i = \Pr\left[d^2(X_i; \hat{\mu}, \hat{\Sigma}) > d^2(x_i; \hat{\mu}, \hat{\Sigma}) \right].$$

Let now α denote the overall level of the test. The FDR is the expected proportion of erroneously rejected hypotheses, if any. In order to guarantee that this expected proportion is below α, it is shown in Cerioli and Farcomeni (2011) that one can use the Benjamini and Hochberg (1995) procedure, at least for large n. One can therefore proceed by sorting the p-values and flagging the observation corresponding to the jth sorted p-value if

$$p_j < j\alpha/n.$$

The FDX is instead defined as the probability that the proportion of false rejections is above a threshold c, commonly set equal to 0.1. It can be controlled at level α as suggested by Lehmann and Romano (2004), by flagging observations corresponding to

$$p_j < (\lfloor jc \rfloor + 1)\alpha/(n + \lfloor jc \rfloor + 1 - j).$$

2.8 Example (Spam detection) *Spam filters are commonly used to block undesired e-mails. A possible approach to set up a spam filter is to define*

TABLE 2.3
Spam data. Number of true and false detected spam messages by RMCD,
FDX-RMCD and FDR-RMCD.

		Detections	
α		True	False
	RMCD	23	421
0.01	FDX-RMCD	11	7
	FDR-RMCD	16	84
0.05	FDX-RMCD	15	61
	FDR-RMCD	20	203
0.10	FDX-RMCD	16	85
	FDR-RMCD	21	326

measurements so that a spam e-mail can be seen as one outlying from the bulk of the other e-mails. These measurements usually involve counts of buzz words which may increase or decrease the suspect of it being undesired.

A primary concern with spam filtering is a control of the rate of swamping. The loss is not symmetric: a false detection may lead to loss of important information, while a false negative is actually negligible as few spam e-mails can be easily identified and deleted by the human reader without the process being too obnoxious. For background on spam see Cranor and La Macchia (1998).

Here we analyze a sample of $n = 2486$ e-mails, thirty of which are spam, and $p = 20$ variables. Data have been pre-processed by adding one unit to each count and then log-transforming. A further standardization step has been carried out by dividing the observations by their column MADs. The processed data set is available from the accompanying website of this book, where the first thirty rows of the data file correspond to spam e-mails.

We use the 50% BP RMCD to estimate the location and covariance structure of the data and the outlier identification rule described in (2.28) and (2.29). In Table 2.3 we report the number of true and false detections based on three methods: the first uses fixed thresholds corresponding to α-quantiles, the second and third are based on an FDR and on an FDX correction, respectively.

It can be seen that using α-quantiles without any adjustment gives the most powerful spam detection rule, but at the price of a large amount of swamping. No e-mail user would ever be prepared to lose so many e-mails. As a matter of fact, α should be calibrated as to have an almost zero swamping in this application. If we did so without using multiple testing corrections, we would incur in a complete masking, that is, in zero detected spam e-mails. The solution is given by multiple testing strategies, where FDX correction seems to be more effective than the FDR correction in this particular case. FDR and FDX both provide a good compromise between masking and swamping. The number of e-mails erroneously marked as spam drops remarkably when controlling these error rates.

2.5 Examples

In this section we illustrate the use of robust multivariate techniques for fitting and outlier identification on some example data sets.

The aim is twofold: on the one hand we want to compare robust solutions with non-robust fits and illustrate the deleterious effects of not taking outliers into account; on the other hand it is also interesting to compare the different robust methods.

In this section we will also introduce the recent approach of *monitoring* as proposed by Riani *et al.* (2014a). Monitoring involves evaluating the robust fit and related quantities (distances and parameter estimates, for instance), as the *breakdown point* or *efficiency* is changed. Monitoring gives more insights into data analyses and allows an empirical selection of tuning constants. In place of making the choice of an appropriate BP and efficiency *a priori*, an investigation of the performance when ranging from high robustness to no robustness at all may be very useful in many applications. The reader is referred to Riani *et al.* (2014a) for details and additional motivating examples and illustrations.

2.5.1 Italian demographics data

The data on natality and mortality rates in 1997 for the 20 Italian regions plot in Figure 1.2 show a clear (negative) relationship between the two variables. Some peculiarities can also be identified: Liguria is the region with the lowest birth rate and the larger death rate and seems to be located quite distant from the bulk of the data, Campania is similarly distant and located in the opposite corner of Figure 1.2 because of its high birth rate. Sardegna looks like a world apart, since it violates the general correlation pattern.

Figure 2.11 shows 97.5%-level tolerance ellipses, based on MLE, the 25% BP RMCD and the MM estimate with 25% BP and 95% shape efficiency. The tolerance ellipses at 97.5% level are expected to contain most of the probability mass, therefore any observation falling outside the ellipse is extreme (it belongs to a region with only 2.5% of the mass) and possibly an outlier.

The tolerance ellipse corresponding to maximum likelihood estimation is based on the distributional result (2.31), the one based on the RMCD involves (2.28), and the one based on MM estimation uses an asymptotic χ_2^2 approximation. Robust fits have a more eccentric shape than the classical one, that is clearly biased by Sardegna. Sardegna is outside the robust tolerance ellipses, whereas Liguria, Campania and Sicilia are close to the boundaries.

Sardegna and Campania both fall in the trimmed set (they receive a null weight) when using the RMCD, and have the two lowest weights, 0.16 and 0.52, respectively, in MM estimation. On the other hand, Liguria and Sicilia receive a unit weight with the RMCD and a slightly larger weight with MM

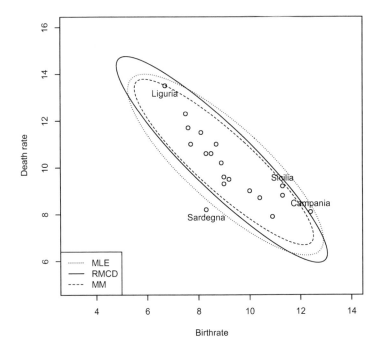

FIGURE 2.11
Italian demographics data. Fitted .975-level tolerance ellipses based on the
MLE (dotted line), the RMCD (solid line) and the MM estimate (dashed
line).

estimation (0.62 and 0.66, respectively). The trimming indicator z_i or the
weight value w_i can always be used as informal proxies for outlier detection.

Formal outlier detection proceeds by computing the robust distances, and
compare them with appropriately chosen cut-off values. The robust distances
from the RMCD (with reweighting based on (2.27)) are displayed in the left
panel of Figure 2.12. The horizontal lines give the .975-level quantile of the
reference distributions (2.28) and (2.29). As outlined previously, these quantiles are different for the data in the trimmed (triangles) and non-trimmed
(circles) set of observations.

The plot undoubtedly confirms the outlying nature of the Sardegna region,
since its robust distance exceeds the reference threshold. The robust distances
for Sicilia, Campania and Liguria are all close to the corresponding reference
lines. They all might be deemed as suspicious, but only Sicilia is formally
flagged, whereas Campania and Liguria are not identified as outliers.

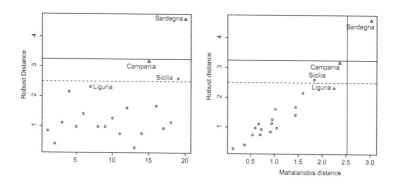

FIGURE 2.12
Italian demographics data. Distance plot (left) and distance-distance plot (right) based on the RMCD.

Another common and useful tool to detect outliers and unveil their effects on classical procedures is the distance-distance plot. The distance-distance plot shows, for each observation, the Mahalanobis distance based on two different estimates.

In the right panel of Figure 2.12 we show the robust distance based on the RMCD versus the Mahalanobis distance based on the MLE for the Italian demographics data.

The vertical line gives the .975-level quantile of the scaled Beta distribution (2.31), whereas the two horizontal lines are based on the approximations in (2.28) and (2.29). Sardegna is confirmed to be an outlier both based on classical and robust methods. As stated previously, Sardegna can be classified as a bad leverage point. On the contrary, the outlying nature of Sicilia is masked in the classical analysis.

Figure 2.13 shows the distance-distance plots based on the χ^2_2 approximation based on the RMCD and MM estimation (vs MLE). Using this approximation Campania is flagged as a clear outlier, whereas Sicilia is not detected. We stress that (2.28) and (2.29) provide a better distributional approximation, hence the previous results are more reliable.

2.5.2   Star cluster CYG OB1 data

The Star Cluster CYG OB1 data set contains measurements about $n = 47$ stars. We are interested in the relationship between the light intensity and the surface temperature ($p = 2$), both measured on the log-scale due to severe skewness. Figure 2.14 reports a scatterplot of the two measures, commonly known as the Hertzsprung-Russell (HR) diagram (the standard HR diagram in astronomy plots the temperatures in reverse order). It can be seen that four

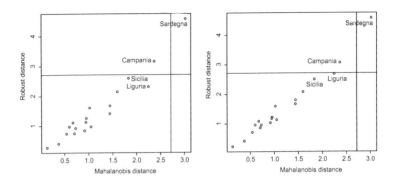

FIGURE 2.13
Italian demographics data. Distance-distance plot based on RMCD (left) and
MM estimation (right). The reference lines are based on the .975 quantile of
the χ^2_2 distribution.

observations clearly stand out and violate the correlation structure: they all
correspond to giant stars (labeled 11, 20, 30, 34).

In the HR diagram we have included the classical, S (with 50% BP), MM
(with 95% shape efficiency) and RMCD (with 50% BP) .975-level tolerance
ellipses. The classical tolerance ellipse is highly biased towards the four giant
stars and is not able to capture the correlation structure of the majority of the
observations. It shows a clearly unrealistic shape. On the contrary, the robust
fits succeed in recovering the true nature of their linear dependence, unveil
the main pattern and clearly identify the four stars belonging to a different
population with respect to the rest of the data. It shall be noted that the four
outliers are known in advance to belong to a different subgroup of stars, so
formally the robust method has been used as a *confirmatory* tool. It should
be additionally noted that while in this (and the previous) example $p = 2$, so
that graphical tools based on scatterplots and tolerance ellipses can be used
to assess contamination, when $p > 2$ it may become difficult to visually assess
contamination based on them. Robust methods are more and more important
as p grows and contamination becomes more complex.

The stars labeled 7,9,14 are also worthy of attention. Figure 2.15 displays
the distance-distance plot based on the MM and RMCD fit. The former is
based on $\chi^2_{.975,2}$ thresholds. The four values corresponding to giant stars are
all clearly identified by both robust fits. Moreover, the stars labeled 7,9,14 are
also detected. These were otherwise masked.

We now focus on S estimation, and monitor what happens to the robust
fit by varying the BP. We start from a value of 50% and decrease this value
by 0.01 up to 0.01. We therefore perform monitoring as suggested by Riani
et al. (2014a). We plot the robust distances for all BP values in Figure 2.16.

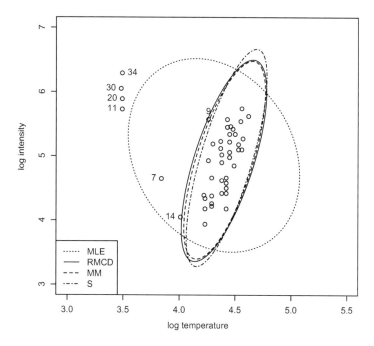

FIGURE 2.14
Stars data. HR diagram with classical (dotted line), MM (dashed line), S
(dotted dashed line), RMCD (solid line) .975-level tolerance ellipses included.

There seem to be three sets of outliers. The set of four outliers (11,20,30,34)
is flagged almost regardless of the BP. Outliers (9, 14) are detected for BP
values larger than 0.34 but they are not identified when the BP is less than
0.27. For BP below 0.14 the robust fit corresponds to maximum likelihood.
We remind that each breakdown value corresponds to a tuning constant c,
according to (2.14) and (2.21). The same constant also leads to a certain
efficiency at the normal model.

By using the monitoring plot in Figure 2.16 it is possible to identify a
minimum BP leading to a robust fit that is *relatively close* to that based
on 50% BP. This can increase efficiency considerably. When $\varepsilon^* = 0.14$ then
$ARE(\hat{\mu}_S, \Phi) = 97.7\%$, which is therefore an upper bound for the location
efficiency. When $\varepsilon^* = 0.35$ then $ARE(\hat{\mu}_S, \Phi) = 80.6\%$ and, finally, when $\varepsilon^* =
0.50$ $ARE(\hat{\mu}_S, \Phi) = 59\%$. In this example we would therefore set $\varepsilon^* = 0.35$,
with a very mild difference in robustness but more than 20% gain in efficiency.
The tolerance ellipses based on S estimation corresponding to each pair (BP,

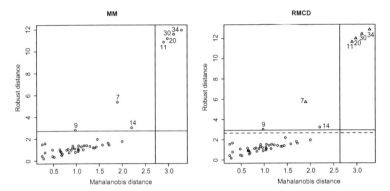

FIGURE 2.15
Stars data: MM (left) and RMCD (right) based distance-distance plot.

ARE) above are shown in Figure 2.17. The classical fit based on the MLE and the S estimate with a BP equal to 0.10 (and ARE equal to 98.9%) are also shown.

Recall that MM estimates directly improve S estimators. Consequently, after monitoring of the breakdown point of S estimators, one could also monitor the efficiency of the resulting multivariate MM estimate. Figure 2.18 shows the robust distances as location efficiency moves between 0.70 and 0.99 (in increments of 0.01). Outlier detection seems to be reliable even for efficiencies close to the upper bound of the grid.

In a similar fashion, Figure 2.19 shows the robust distances based on RMCD as a function of BP: the three sets of outliers are visible even for small breakdown values.

2.9 R Illustration (Robust estimation of the Stars data) *The robust fits plot in Figure 2.14 were computed with the following* R *code. The data is shipped within the* rrcov *package. More details on this library and the functions it supplies are given in Appendix A. The data were loaded and estimates computed as:*

```
> library(rrcov)
> data(starsCYG)
> source("CovMcdF.R")
> rmcd.stars<-CovMcdF(starsCYG, alpha = 0.5)
> mm.stars<-CovMMest(starsCYG,bdp=0.5,
+ eff=0.95, eff.shape=TRUE)
```

The plots involving the MLE and the RMCD are based on the scaled Beta and scaled F rather than on the standard χ_2^2 distribution. Therefore, the standard and RMCD tolerance ellipses were obtained with an ad-hoc R *code. The scaled Beta reference quantiles were computed as*

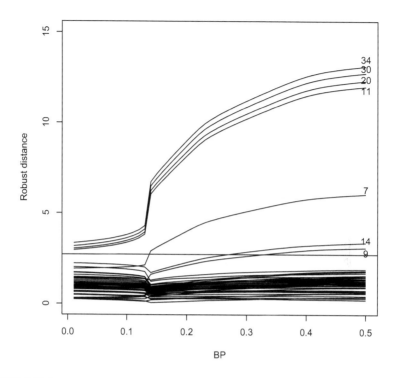

FIGURE 2.16
Stars data. Monitoring the robust distances from S estimation for varying BP values. The horizontal dashed line gives the square root of the .975 quantile of the χ^2_2 distribution.

```
> c=((n-1)^2)/n
> q.mle<-qbeta(.975,p/2, (n-p-1)/2)*c
> w<-sum(rmcd.stars$wt)
> cw=((w-1)^2)/w
> q.rmcd<-qbeta(.975,p/2, (w-p-1)/2)*cw
```

for the MLE and RMCD, respectively. Then, the plot was obtained with

```
> library(ellipse)
> hmu<-colMeans(starsCYG)
> S<-cov(starsCYG)
> plot(starsCYG)
> lines(ellipse(x=S, centre=hmu, t=sqrt(q.mle)),lty=3)
> lines(ellipse(x=rmcd.stars$cov
+ centre=rmcd.stars$center, t=sqrt(q.rmcd)),lty=1)
```

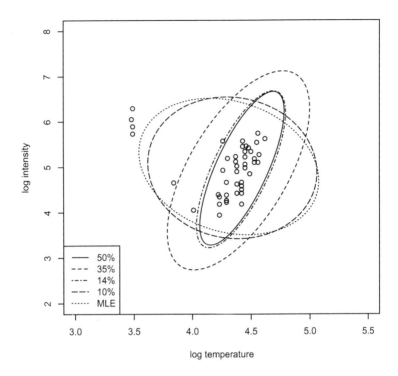

FIGURE 2.17
Stars data. Different fits based on S estimation for different BP values.

We now focus on the distance-distance plot in the right panel of Figure 2.15. First, the cut-off value that allows to detect outliers in the trimmed set of observations from the RMCD need to be evaluated.

```
> cw2=(w^2-1)*p/w/(w-p)
> q.out<-qf(.975,p, w-p)*cw2
```

The Mahalanobis distances based on $(\bar{x}, S)$ are

```
> mle.mah<-mahalanobis(starsCYG, hmu,S )
```

Then, the plot is obtained through the following code:

```
> plot(sqrt(m.stars), sqrt(rmcd.stars$mah),type='n')
> abline(h=c(q.out, q.rmcd),lty=1:2)
> abline(v=q.mle)
> inx.trim<-which(rmcd.stars$wt==0)
> inx.non.trim<-which(rmcd.stars$wt!=0)
```

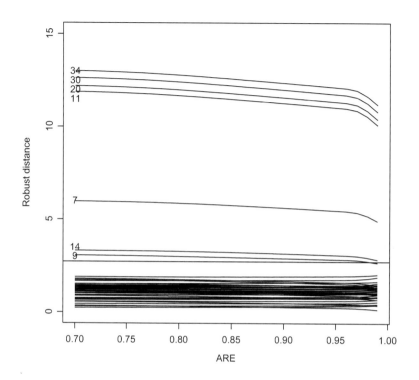

FIGURE 2.18
Stars data. Monitoring the robust distances from MM estimation for $\varepsilon^* = 50\%$ and varying location efficiency values. The horizontal dashed line gives the square root of the .975 quantile of the χ_2^2 distribution.

```
> points(sqrt(mle.mah[inx.non.trim]),
+ sqrt(rmcd.stars$mah[inx.non.trim]),pch=1)
> points(sqrt(m.stars[inx.trim]),
+ sqrt(rmcd.stars$mah[inx.trim]),pch=2)
```

If one is willing to use the χ_2^2 approximation, tolerance ellipses, distance-distance plot and other graphical tools for outlier detection can be simply accessed through the function `plot` applied to the outcome of the `CovMMest` function. The option `which` allows to select the desired plot. This feature will be illustrated in detail in Section A.1. The reader is also pointed to Todorov and Filzmoser (2009) for an extensive survey of the `rrcov` library.

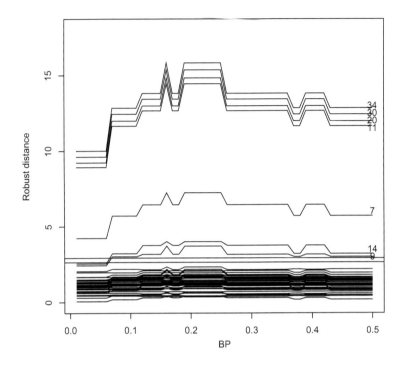

FIGURE 2.19
Stars data. Monitoring the robust distances from RMCD for varying BP values. The horizontal dashed lines gives the reference quantile based on the scaled Beta (dashed) and scaled F (dotted) distribution.

2.5.3 Butterfly data

Let us consider the butterfly data of Section 1.4. These data concern four environmental attributes for 16 colonies of *Euphydryas edhita*, observed in California and Oregon. We already mentioned that two colonies, namely GH and GL, are anomalous and badly influential for the correlations.

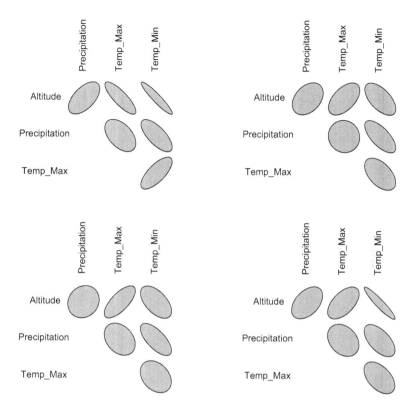

FIGURE 2.20

Butterfly data. Pairwise correlations in the form of ellipses. Clockwise: classical, RMCD, OGK, SD.

Here we compare the results based on RMCD, OGK and SD estimators. In the four panels of Figure 2.20 we report the estimated correlation structure, where estimates are based on the standard approach and on the three robust estimators. Each plot can be replicated using function `plotcorr` in package `ellipse`.

Anomalous altitudes and temperatures lead to estimate a reverse linear relationship between `altitude` and `maximum temperature` and between `maximum temperature` and `minimum temperature` when robust methods are not used.

Figure 2.21 displays the distance plots based on the three robust methods. The first two distance plots (top row) are based on cut-off values as in Sections 2.3.5 and 2.4, the other two (bottom row) are based on a χ^2_4 approximation. The classical method is clearly unreliable due to disagreement with the robust counterparts. The two anomalous colonies `GH` and `GL` are not detected using the classical Mahalanobis distances. On the contrary, they are clearly identified

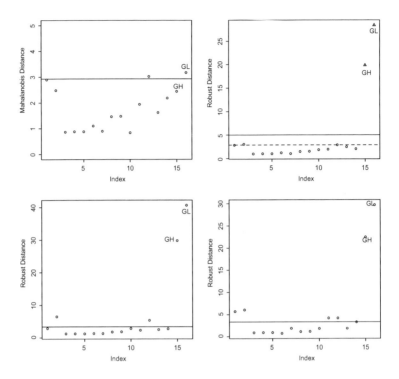

FIGURE 2.21
Buttefly data. Distance plots based on MLE, RMCD, OGK and SD estimates.

as outliers by the robust methods. By using the sample mean and covariance, only the colony GH is slightly above the threshold line but GH lies below. Furthermore, it is worth noting that none of the points would have been detected after using the standard approach if the χ_4^2 approximation were used in place of the exact scaled Beta distribution.

Part I

Dimension Reduction

Introduction to Dimension Reduction

In some applications in which a large number of variables p is recorded, an essential step consists in obtaining a low dimensional representation of the data. This projection should be as faithful as possible to the original measurements, that is, it should retain as much of the original information as possible.

The main idea is that a high dimensional problem can actually be simplified and described in terms of fewer measurements. The new measurements are functions of the original ones. These functions are chosen so as to maximize agreement between original and new measurements.

In some cases not all variables are relevant for understanding the phenomena of interest and many of them could be highly correlated (hence redundant). In this case dimension reduction might be successful in substantially compressing information.

Furthermore, projection onto a two or three dimensional space allows visualization of multivariate samples, and hence improves comprehension of the underlying structure. Bidimensional graphical displays (sometimes referenced as *biplots* or *score plots*) may unveil important features otherwise masked in high dimensions. It should be kept in mind, in any case, that also the contrary can happen as projections may hide complex features and even outliers.

Dimension reduction additionally simplifies storing of huge amounts of data by compressing to a manageable size.

Principal Component Analysis (PCA) is one of the most widespread techniques for dimension reduction. There is a plethora of applications in many fields: image and video processing, speech processing, chemometrics, engineering, astronomy, biology, genomics, economics, finance, only to mention some. PCA reduces the dimension of the data by finding *orthogonal* linear combinations of the original variables. Each linear combination is a new variable. Additionally, each new variable should have the largest variance possible. The new variables can be interpreted and labeled for further use and allow graphical displays.

The main techniques for robust PCA are discussed in Chapter 3, whereas in Chapter 4 the attention is focused on sparse PCA. We then proceed with other well known methods for dimension reduction: Canonical Correlation Analysis in Chapter 5 and Factor Analysis in Chapter 6.

For an in-depth understanding of the methods in the first part of this book, the material of Chapter 2 is a prerequisite.

3

Principal Component Analysis

This chapter is an introduction to the field of robust multivariate techniques for dimension reduction. We focus on principal components analysis, and its robust counterparts.

In the first section of this chapter we review classical PCA. PCA is a widely used technique for descriptive multivariate statistics and dimensionality reduction. This material will only be covered shortly, in order to keep the book self contained. The reader can refer to Jolliffe (2005) for an extensive account on classical dimension reduction.

There have been many efforts to obtain robust versions of PCA. A first group of strategies is based on computing eigenvalues and eigenvectors based on a robust estimate of the covariance matrix (Croux and Haesbroeck, 2000; Salibian-Barrera *et al.*, 2006). This works well as long as robust estimation of multivariate location and covariance is feasible (e.g., when p is small enough).

A different approach is based on projection pursuit (Croux and Ruiz-Gazen, 2005; Croux *et al.*, 2007). These methods are well suited to handle high dimensional data and situations in which the number of variables exceeds the number of observations. Actually, in these cases robust covariance estimation methods become troublesome or even impossible to perform. In projection pursuit, principal components are found in sequence, without the need to estimate the covariance matrix. This is achieved by maximizing a robust measure of variability instead of the sample variance. A related approach is that of Hubert *et al.* (2005): ROBPCA combines projection pursuit with robust estimation of the covariance matrix. There are also other proposals for robust PCA, among those we mention spherical PCA introduced by Locantore *et al.* (1999) and orthogonal PCA developed by Maronna (2005).

All the approaches to robust PCA listed above have been conceived to handle structural outliers. We will briefly mention also the case of component-wise outliers throughout the book. To cope with sparse contaminated entries, *alternating regression* schemes have been proposed in the literature. These are not covered in this book: De La Torre and Black (2003) proposed element-wise M estimates, Maronna and Yohai (2008) suggested the use of MM estimates, whereas Croux *et al.* (2003) developed a weighted L_1 robust alternating regression (RAR) scheme.

Finally, robust covariance estimates like 2GSE (Agostinelli *et al.*, 2014), snipEM or stEM (Farcomeni, 2014a, 2015) can be used to directly implement robust PCA methods resistant to component-wise and structural outliers. Brief notes on the implementation of the last two methods are given in appendix.

Robust PCA can also be a preliminary step before the identification of multivariate outliers. This is a very attractive feature of robust PCA. Identification of outliers can be a very difficult task in high dimensions. Outliers may be more easily identified after reducing dimensionality. Finding the (robust) principal components is often a preliminary step in other contexts of data analysis, especially when there are many variables. PCA can be followed by regression, classification, cluster analysis and other multivariate techniques (Hubert and Engelen, 2004).

3.1   Classical PCA

PCA aims at reducing the number of variables to a more manageable size by projection onto a lower dimensional space. The final representation of information is simpler not only because of lower dimensionality, but also because one obtains new variables that are orthogonal and have zero correlation. Hence the information is separated into independent nuggets. An additional feature is that the projection is performed so that observations are as spread out as possible. The underlying idea is that information *is* variability. This statement has an intuitive interpretation when one thinks about zero variance. A variable with zero variance does not give any information, as observations are all equivalent according to that dimension.

The original p-dimensional set of variables $\mathbf{X}$ is mapped to a new set of orthogonal and uncorrelated variables $\mathbf{Y} = \left(Y_{(1)}, Y_{(2)}, \ldots, Y_{(p)}\right)^{\mathrm{T}}$, (the *principal components* or *scores*). These new variables are obtained as linear combinations of the initial ones. According to the coefficients of the linear combination, each p-dimensional observation is transformed into a *score*. The scores are collected in an n by p matrix Y. A very useful property of PCA is that for each $q \geq 1$, the set of q variables that are orthogonal, uncorrelated and with the least information lost is given by the first q columns of Y. In this way, the best summary in $q = 2$ dimensions is given by the first two scores, the best summary in $q = 5$ by the first five and so on. An additional advantage is that PCA has to be computed only once, obtaining all the possible solutions up to $q = p$ simultaneously.

For simplicity we assume in the following that data are zero centered, that is, column sample means $\bar{x}$ have been previously subtracted. The first principal component (PC) is obtained by searching for the linear combination

$$Y_{(1)} = a_1^{\mathrm{T}}\mathbf{X} = a_{11}X_1 + a_{12}X_2 + \ldots + a_{1p}X_p$$

whose sample variance $s_1^2 = a_1^{\mathrm{T}} S a_1$ is the largest among all possible linear combinations of the original variables, under the constraint of unit norm $||a_1|| = 1$. The constraint is needed because, otherwise, one can pick any diverging a_{1j} and have an unbounded variance. Note that with $Y_{(1)}$ in this chapter we identify the first column of the scores matrix Y, and implicitly underline that it contains the largest possible variability. Similarly, we will use notation $Y_{(j)}$ to denote the jth column of Y. The vector a_1 that solves the constrained optimization problem is the first eigenvector ν_1 of the sample covariance matrix S. The variance of $Y_{(1)}$ is equal to the first sample eigenvalue λ_1. The elements of ν_1 are called *loadings* of the first PC. They not only lead to the first score, but can be also interpreted as a measure of relative importance of each variable for the component. A loading far from zero corresponding to X_j indicates that the jth original variable has a large importance for $Y_{(1)}$. It *characterizes* the first score. This is an important note for interpretation (labeling) of the scores. The scores can be thought as (signed) weighted averages and PCA can also be thought as an automatic summary index building device. The scores are indicators, but need *post hoc* interpretation based on the loadings.

The other PCs are computed in a similar fashion by maximizing the same criterion but under the additional constraint that they are orthogonal to the previous ones, that is $Y_{(j)}^T Y_{(i)} = 0, j < i$. This is precisely a property of eigenvectors. It can be shown, in fact, that the jth PC is given by $Y_{(j)} = \nu_j^T \mathbf{X}$, where ν_j is the jth eigenvector of S. The eigenvectors are ordered with respect to the magnitude of their eigenvalues: given S must be positive definite, we have

$$\lambda_1 \geq \lambda_2 \geq \ldots \geq \lambda_p > 0.$$

In matrix notation, the score matrix can be obtained as $Y = XV^T$, where the columns of V are the eigenvectors ν_j, with $j = 1, 2, \ldots, p$. Since $V^T V = I_p$, where I_p is the identity matrix of dimension p, from the orthogonal decomposition of the sample covariance matrix $S = V\Lambda V^T$, with $\Lambda = \text{diag}(\lambda_j)$, it is straightforward to see that the covariance structure of the principal components is given by $S_Y = \Lambda$. Proof of this is given by the following equations:

$$S_Y = Y^T Y = V^T S V = V^T V \Lambda V^T V = \Lambda .$$

It is worth mentioning that PCA can be also carried out by singular value decomposition (SVD) of the centered data matrix $X - \bar{x}$

$$X - \bar{x} = U\Lambda V^T = YV^T .$$

Given the properties that the maximal amount of information that can be retained with q linear combinations is contained in the first q columns of Y, we can discard the remaining last $p - q$ columns of Y to reduce dimensionality in an optimal way. The number $q < p$ of principal components to retain is usually chosen on the basis of the proportion of *explained variance*. The total variance of the original variables is equal to the trace of S, that is $\text{tr}(S) = \sum_{j=1}^{p} S_{jj}$, and equivalently to the sum of the eigenvalues of S. The trace can in fact be expressed as a function of the eigenvalues:

$$\text{tr}(S) = \text{tr}(V \Lambda V^{\text{T}}) = \text{tr}(V^{\text{T}} V \Lambda) = \sum_{j=1}^{p} \lambda_j = \text{tr}(S_Y) \ .$$

It follows that the total variance of the original variables equals the total variance of the p principal components. Then, the percentage of total sample variance explained by the first $q < p$ principal components is

$$\frac{\sum_{j=1}^{q} \lambda_j}{\sum_{j=1}^{p} \lambda_j}.$$

For instance, one may want to select q so that 70% or 80% of the original variability is explained. Another very popular criterion is based on the Kaiser's rule, according to which only the components that have eigenvalues greater than the average variance are retained for interpretation (when using the correlation matrix one could select components whose eigenvalues are greater than one). For a general account on rules for selecting q see Zwick and Velicer (1986).

When a large proportion of the total variance is explained by the first q components, one can ignore the original p variables and focus on the PCs without much loss of information. Let us call Y_q the $(n \times q)$ matrix whose columns contain the scores on the first q principal components, given by

$$Y_q = X V_q$$

where $V_q = [\nu_1, \nu_2, \ldots, \nu_q]$ collects the first q eigenvectors of S. It can be noted that the new data matrix Y_q can be used also to obtain an approximation to the original data matrix X as follows:

$$\tilde{X} = Y_q V_q^{\text{T}} + \bar{x}.$$

When n and p are large, V_q, Y_q and $\bar{x}$ can be stored much efficiently than X, therefore obtaining information compression. The first q PCs also provide a rank q approximation of S, namely

$$S_q = V_q \Lambda_q V_q^{\text{T}} \ .$$

Besides being an effective compression technique, PCA is a powerful tool for feature extraction and feature selection. The loadings of the selected components can help to unveil information, as measured by the original p variables and their associations. After interpretation the components might be labeled for further use. Feature selection can also be approximately performed, as the most important variables will usually be associated with loadings that are far from zero. Variables receiving loadings close to zero in V_q are often discarded.

TABLE 3.1
Spam data. Percentage of variance explained by PCA.

q	2	3	5	8	10
%	33.3	40.4	51.8.	65.5	73.3

In situations in which the variables are measured on different scales and the variances differ widely, one may wish to use the sample correlation matrix instead of the sample covariance matrix. Otherwise, variables with the larger variances will dominate the first components regardless of their true importance.

The main R function to perform classical PCA is `princomp`, whereas the function `eigen` allows to compute eigenvalues and eigenvectors of a positive definite square matrix and `svd` performs singular value decomposition. The function `screeplot` provides a graphical display of the percentage of explained variance from each component. Statistical folklore is that an elbow in the screeplot corresponds to the optimal q.

3.1 Example (Dimension reduction of spam data) *Let us consider the data relative to spam detection. It would be interesting to reduce the number of attributes which could indicate an e-mail being clean or spam. PCA is based on the sample covariance matrix. In this example variables are of very similar nature and have the same unit of measurement. Therefore it is not necessary to standardize.*

In Table 3.1 we show the percentage of explained cumulative variance attained from the first q components. It can be noted that $q = 5$ components are able to explain almost half of the initial variability, whereas $q = 10$ leads to 73.3% explained variance. The screeplot in Figure 3.1 indicates an elbow at $q = 3$: the first three components alone are able to explain 40.4% of the total variance, after the fourth we have slower increases in variability. Figure 3.2 shows the loadings of the first three components. Some of them are very large, other are not. There are also variables, namely $(2, 3, 4, 12, 16, 20)$, whose importance is negligible and could be discarded when computing the first three components.

3.2 R Illustration (PCA of spam data) *Classical PCA was performed via the following code.*

```
> pca.spam<-princomp(spam20)
> summary(pca.spam)
```

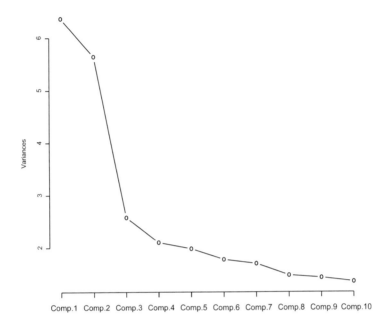

FIGURE 3.1
Spam data. Screeplot based on classical PCA.

Figures 3.1 and 3.2 were obtained as follows

```
> screeplot(pca.spam,npcs=10,type='lines',main='')
> barplot(t(pca.spam$loadings[,1:3]),
+ beside=TRUE,legend.text=TRUE)
```

3.2   PCA based on robust covariance estimation

Since PCA is based on the sample covariance matrix, anomalous values may lead to completely biased and unreliable results: it can be shown that a single outlier can breakdown the procedure completely. Under contamination, it often happens that the first PC is *shifted* towards outliers. Outliers in fact lead to inflated variability towards their direction. As a consequence, one ends with overly optimistic eigenvalues and, more importantly, with loadings that

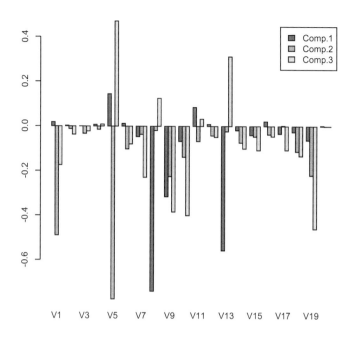

FIGURE 3.2

Spam data. Loadings of the first three components.

are rotated with respect to the direction of largest variability pertaining to the bulk of the data. It follows that the interpretation of the components may be determined solely by outliers and relevant features of the clean part of the data may be hidden by the projection.

These problems can be overcome by using robust methods for PCA.

A simple and appealing approach consists in deriving the principal components based on a robust (affine equivariant) estimate of the covariance (correlation) matrix. One can simply use the the S estimator, the MM estimator (Salibian-Barrera *et al.*, 2006), or the RMCD estimator (Croux and Haesbroeck, 2000) and then obtain eigenvalues and eigenvectors from the robust estimate.

The corresponding *robust* multivariate location estimate is used to center the data and to approximate the original data matrix by means of the first q robust components.

The robustness properties of PCA based on a robust estimate of the covariance matrix can be explored by looking at the influence functions for eigenvectors and eigenvalues: the reader is pointed to Croux and Haesbroeck

(2000) and Salibian-Barrera *et al.* (2006) for the technical details. One can also use sensitivity curves, which are simple to implement.

A different robust PCA solution is obtained for each different robust estimate. The results based on different robust techniques must be compared. In alternative, the choice of the robust covariance estimator used may be justified by the characteristics of the data.

Robust PCA is handled directly by the R function `PcaCov` in library `rrcov`. Remember anyway that RMCD in library `rrcov` is based on the classical χ^2_p approximation and that efficiency of the MM estimator can be tuned only with respect to location in a PCA problem. The basic function `eigen` can be used as an alternative if one wishes to work with a different robust estimate of the covariance matrix, for instance, with an RMCD based on the reweighing scheme of Hardin and Rocke (2004), discussed in Section 2.3.5.

The robust PCA methods outlined here are extremely simple and their principle is absolutely equivalent to that of classical PCA. Unfortunately, they are limited to small or moderate dimensions. In high dimensions robust estimates may not be feasible. The fastest algorithms can handle up to $p = 100$, but they work only when $n >> p$. A larger sample size than the number of variables is needed in order to guarantee a positive definite covariance matrix.

3.3   PCA based on projection pursuit

An approach to robust PCA based on projection-pursuit (PP) is proposed in Li and Chen (1985). PP methods are specifically aimed at projecting multivariate data onto a lower dimensional subspace. The lower dimensional subspace is found by maximizing a *projection index*. This methodology parallels the PCA principle of searching for directions with maximal variance.

If we use the classical variance as projection index, then classical PCA is recovered. A robust PP method can be defined by replacing the sample variance by a robust measure of variability.

The advantage of PP is that one does not need estimating the covariance matrix. Robust estimates of the eigenvalues and the eigenvectors are obtained sequentially. This makes PP applicable also when p is very large and $p > n$.

Assume that the data have been previously centered by using the column-wise median or the spatial median (or L_1 median, Hössjer and Croux, 1995; Croux and Ruiz-Gazen, 2005):

$$Med(X)_{L_1} = \mathrm{argmin}_{\mu \in \mathbb{R}^p} \sum_{i=1}^{n} ||x_i - \mu||\,,$$

where $||x_i||^2 = \sum_{j=1}^{p} x_{ij}^2$ denotes the squared norm of the p-dimensional vector x_i. Let $\hat{\sigma}$ denote a robust univariate scale estimator. Different choices can be made about the robust measure of dispersion to be maximized: the MAD, the Q_n estimator of Rousseeuw and Croux (1993), a scale M estimator, an L-scale, such as the trimmed squared scale estimate (Rousseeuw and Leroy, 1987). The first loading vector is obtained as

$$\nu_{\hat{\sigma};1} = \text{argmax}_{||a||=1}\hat{\sigma}^2(a^\mathrm{T}x_1, a^\mathrm{T}x_2, \ldots, a^\mathrm{T}x_n) \qquad (3.1)$$

with associated eigenvalue

$$\lambda_{\hat{\sigma};1} = \hat{\sigma}^2(\nu_{\hat{\sigma};1}^\mathrm{T}x_1, \nu_{\hat{\sigma};1}^\mathrm{T}x_2, \ldots, \nu_{\hat{\sigma};1}^\mathrm{T}x_n).$$

This directly mimics the classical PCA projection representation.

The qth PC is defined by

$$\nu_{\hat{\sigma};q} = \text{argmax}_{||a||=1}\hat{\sigma}^2(a^\mathrm{T}x_1, a^\mathrm{T}x_2, \ldots, a^\mathrm{T}x_n) \qquad (3.2)$$

under the further constraint of orthogonality of the eigenvectors with corresponding eigenvalue

$$\lambda_{\hat{\sigma};j} = \hat{\sigma}^2(\nu_{\hat{\sigma};j}^\mathrm{T}x_1, \nu_{\hat{\sigma};j}^\mathrm{T}x_2, \ldots, \nu_{\hat{\sigma};j}^\mathrm{T}x_n), j = 1, 2, \ldots, q.$$

We notice that the variability of the component is measured directly by a robust estimate of the variance of the scores, whereas in PCA based on a robust estimate of the covariance matrix the variability of the component is accounted for by the corresponding eigenvalue.

Distributional results and efficient algorithms for solving (3.1) and (3.2) have been proposed by Hubert *et al.* (2002), Croux and Ruiz-Gazen (2005) and Croux *et al.* (2007). Robust PCA based on PP is available from the functions PCAproj and PCAgrid in library pcaPP, but also from the function PcaProj in library rrcov.

We stress that a different robust PCA method, with corresponding solution, is obtained for each choice of the robust projection index. The MAD and the Q_n estimator are possible choices. Our suggestion is to compute always and compare more than one robust estimate to check sensitivity to the projection index used. An understanding of the properties of each robust scale estimator may help in interpreting the results and choosing the most reliable.

A robust PCA method that is similar to projection pursuit has been proposed in Maronna (2005).

3.4 Spherical PCA

A further approach to robust PCA has been introduced in Locantore *et al.* (1999) and is called *spherical* PCA (SPCA). The idea is to perform classical

PCA of data projected onto the surface of a unit sphere to reduce the effect of outliers. Assume the data have been previously centered by a robust multivariate location estimate. Let us consider the transformed data

$$\tilde{x}_i = \frac{x_i}{||x_i||} \,.$$

Each row of the (centered) data matrix X is divided by its norm. Let $\tilde{X}$ be the transformed data matrix whose i^{th} row is $\tilde{x}_i$. Spherical PCA is based on the eigenvectors $\nu_{sph;j}$ of the covariance matrix of $\tilde{X}$. Summarizing, one should first project the data, then compute the classical covariance (or correlation) matrix out of $\tilde{X}$, and proceed with standard PCA.

The idea behind this procedure is that when outliers are not present, the data approximately should lie close to their projection. This is testified by the fact that, in absence of contamination, the eigenvectors of the projected covariance matrix and the eigenvectors of the original data matrix get closer and closer as the sample size grows. As well as in PCA based on PP, eigenvalues are computed through a robust measure of scale, that is

$$\lambda_{sph;j} = \hat{\sigma}^2(\nu_{sph;j}^{\mathrm{T}} x_1, \nu_{sph;j}^{\mathrm{T}} x_2, \dots, \nu_{sph;j}^{\mathrm{T}} x_n), j = 1, 2, \dots, q.$$

The robust measure of scale might be the MAD or the Q_n estimator. The function `PCaLocantore` from library `rrcov` can be used to perform SPCA.

3.5   PCA in high dimensions

The PP and SPCA approaches to robust PCA are feasible in high dimensions. A method that has been explicitly conceived to directly handle high dimensional and $p > n$ data is ROBPCA of Hubert *et al.* (2005) and Engelen *et al.* (2005). This technique combines ideas from projection pursuit and robust covariance estimation based on the RMCD. ROBPCA proceeds as follows

1. First, the data space is projected to a space whose dimension r_0 is at most $(n - 1)$. This dimensionality reduction is achieved by singular value decomposition of the centered data matrix $X_{n \times p} - \bar{x}$. One obtains $Y_{n \times r_0}$ and $V_{p \times r_0}$. This is also possible when $p > n$.

2. Let $Y_{n \times r_0}$ be the new data matrix. Then, the Stahel-Donoho method (2.30), based on the couple (median, MAD) or on the univariate MCD, is used to identify outliers. The sample size is then reduced to $n(1-\varepsilon) \leq n$. If *exact fit* is achieved, characterized by a null robust scale for some dimensions, the initial dimensionality r_0 is reduced to $r_1 < r_0$.

3. The sample covariance S_0 (and location $\hat{\mu}_0$) of the h clean observations is computed and used to decide how many components q_0 to

retain. In Engelen *et al.* (2005) it is advised to work with a reweighed version of S_0 obtained by assigning a null weight to those observations with a large distance from the subspace spanned by the first q_0 eigenvectors of S_0, collected in the matrix $V_{r_1 \times q_0}$.

4. The data in $Y_{n \times r_1}$ are projected onto the subspace spanned by S_0 as

$$X^*_{n \times q_0} = (Y_{n \times r_1} - \hat{\mu}_0)V_{r_1 \times q_0}.$$

In order to avoid replication of the technique for several choices of q, hence leading to a very time consuming approach, Engelen *et al.* (2005) suggested to project the data onto a q_{max}-dimensional subspace, and then find the optimal q_0. This involves performing this step *before* the previous one.

5. The covariance of $X^*_{n \times q_0}$ is estimated through the RMCD, yielding $\hat{\Sigma}_1$ whose dimension is $q \times q$, with $q \leq q_0$, since *exact fits* may occur also at this step.

6. The final PCs are obtained through classical PCA on the robust estimate $\hat{\Sigma}_1$ and by transforming the eigenvectors of $\hat{\Sigma}_1$ that lie in a q dimensional subspace back to $\mathbb{R}^p$. A final rank-q p-dimensional estimate of the covariance is obtained as

$$\hat{\Sigma} = V_{p \times q} D_{q \times q} V^{\mathrm{T}}_{p \times q},$$

where D is the diagonal matrix whose elements are given by the eigenvalues of $\hat{\Sigma}_1$.

The robustness properties of ROBPCA and its asymptotic behavior have been investigated in Debruyne and Hubert (2009). ROBPCA can be performed in R with the function `PCAHubert` from library `rrcov`.

3.6   Outlier identification using principal components

Dimensionality reduction through robust PCA can greatly increase the accuracy of outlier identification methods.

Let us focus on the PCA space spanned by the first q components. Outliers can be identified by measuring the distance of each projected data point $y_i = (y_{i1}, y_{i2}, \ldots, y_{iq})^{\mathrm{T}}$ from the origin. This is called a *score distance* and is given by

$$SD(y_i; \Lambda_q) = \sqrt{y_i^{\mathrm{T}} \Lambda_q^{-1} y_i}, \; i = 1, 2, \ldots, n. \tag{3.3}$$

A point may be flagged when a large score distance is observed. Cut-off points for outlying squared score distances are determined by using the quantiles of a χ_q^2 distribution. This approach clearly suffers from the same problems discussed in Chapter 2 concerning the accuracy of the approximation to the distribution of robust distances. Moreover, multiplicity issues should be taken into account.

Let $\tilde{X} = Y_q V_q^{\mathrm{T}}$ be the approximation to the centered data matrix by means of the first q PCs, stored in Y_q. The distance of each point x_i to the q-dimensional subspace gives an idea of the accuracy of the approximation. It is called *orthogonal distance* and is given by

$$OD(x_i; \tilde{x}_i) = ||x_i - \tilde{x}_i||, \ i = 1, 2, \ldots, n, \tag{3.4}$$

where $\tilde{x}_i$ is the ith row of $\tilde{X}$. In order to compute appropriate cut-off values, we refer to Hubert *et al.* (2005) and Hubert *et al.* (2008), where the Wilson-Hilferty approximation is used. According to these results, the orthogonal distances, raised to the power $2/3$, are approximately normally distributed. Then, thresholds can be found as $(\hat{\mu} + z_\alpha \hat{\sigma})^{3/2}$, where $(\hat{\mu}, \hat{\sigma})$ are robust univariate estimates of location and scale evaluated on $OD(x_i; \tilde{x}_i)^{2/3}$ and z_α the α quantile of the standard normal distribution.

By looking at both score and orthogonal distances, we may obtain different types of possible outliers. There could be points whose projection onto the q-dimensional manifold lies far from the bulk of the projected data. Other points might be suspiciously far from the PCA subspace, but otherwise undetectable by looking only at their projection. Some points might be simultaneously distant from the majority of the projected data and the PCA subspace. In order to identify the different type of outliers, both rules outlined above must be used. The points with a large score distance but small orthogonal distance are *good leverage points*: they are approximated well, therefore being leverage points, but they are far from the origin therefore being *extreme* in some sense. The points with a large orthogonal distance but small score distance are called *orthogonal outliers*: they lie within the bulk of the data, but their approximation is not good, meaning their position is rather awkward with respect to the orientation of the data cloud. The points with both large score and orthogonal distance are *bad leverage points*, or influential outliers: they are both extreme and badly approximated, meaning that they may bias dangerously the standard projection. A powerful graphical tool to distinguish among types of outliers is given by the the outlier map described in Hubert *et al.* (2005) and illustrated in Figure 3.3 for the Butterfly data introduced in Section 1.4. For each observation the score distance (3.3) and the orthogonal distance (3.4) are plot on the horizontal and vertical axes, respectively. The cut-off values divide the graphical display in four quadrants: regular observations are in the bottom-left quadrant and clockwise we find orthogonal outliers, bad leverage points and good leverage points. According to this rule, we have one orthogonal outlier and two bad leverage points for these data. The SB colony is not an outlier *per se*, but it is not approximated well after projection onto a two dimensional space. The other two colonies were already discussed as being influential outliers, and the plot confirms this.

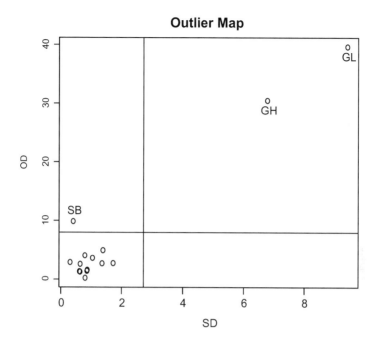

FIGURE 3.3
Butterfly data. Outlier map based on the first two components and RMCD.

3.7 Examples

The following real data examples are used for an illustration of the different robust PCA techniques, interpretation of the results and evaluation of the benefits of robust approaches.

3.7.1 Automobile data

The 1985 Automobile Imports Database is related to the investigation of the main risk factors with insurance policies (Kibler *et al.*, 1989). In this section we will explore the main features of the data through PCA.

In order to perform the analysis, we work with the complete data matrix (after discarding observations with missing values). We have $n = 195$ observations and $p = 15$ quantitative variables. A different version of the same data set is available from the R library `rrcovHD`.

The data include twenty cars that are equipped with a diesel engine. We keep them for illustration, as they could be identified as outliers with respect to the majority of cars running on gasoline. A first look at the data

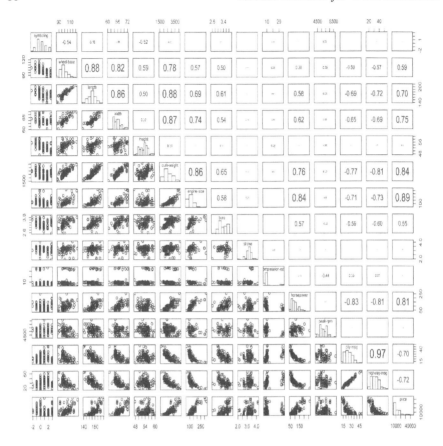

FIGURE 3.4
Automobile data. Scatterplot between all pairs of variables and related correlation coefficients.

in Figure 3.4 reveals that some pairs of variables are highly correlated. The strong correlations are a clear indication that dimensionality reduction might be successful. Moreover, some of the pairwise scatterplots display points that deviate from the majority of the data. For instance, panels corresponding to the variable `compression ratio` highlight the existence of the two groups of observations: cars running on diesel have larger compression ratios.

Figure 3.5 gives classical and robust screeplots. Robust screeplots are slightly more peaked. The percentage of cumulative explained (robust) variance is also given in the last line of Table 3.2 for the first three components.

3.3 R Illustration (Robust PCA of automobile data) *Robust PCA based on the RMCD and MM estimation can be obtained with the following code.*

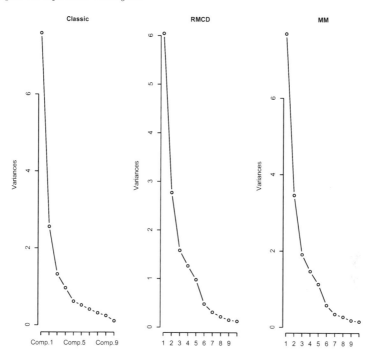

FIGURE 3.5
Automobile data. Screeplot based on classical and robust PCA.

*More details on the options available are given in Appendix A.2. The data have been pre-processed by dividing each column by the corresponding MAD, so all the variables have comparable variability. The second column records the type of engine (*gasoline/diesel*) and is here discarded. The* getLoadings *function from library* rrcov *is used to extract the loadings.*

```
> X<-Automobile[,-2]
> xmad <- apply(X, 2, mad)
> mcd.acp<-PcaCov(X, scale=xmad,
+ cov.control=CovControlMcd(alpha=0.75))
> getLoadings(mcd.acp)
> mm.acp<-PcaCov(X, scale=xmad,
+ cov.control=CovControlMMest(bdp=0.5,eff=0.95))
> getLoadings(mm.acp)
```

In Table 3.2 we report the loadings of the first three components from classical PCA and PCA based on the 25% breakdown point RMCD; and the 50% breakdown point and 95% location-efficient MM estimate of the covariance matrix. The analysis has been performed by using the function `PcaCov` for robust PCA from the `rrcov` library, with the default options. Therefore, they are immediately reproducible. The RMCD is based on the classical χ_p^2 approximation and the efficiency of the MM estimate can be controlled only with respect to location, while the shape efficiency is left free. Any different choice of BP or efficiency can be specified with the `cov.control` argument.

If we look at the first PC we see a slight difference between robust and classical methods. The signs are reversed, which is not important. On the other hand, `length` seems to be slightly more important and `curb-weight`, `engine-size` and `bore` slightly less. As expected, `compression-ratio` has a completely different loading when comparing robust and classical estimates. This is due to the fact that outliers are mostly different with respect to this measure. The variability accounted for by the first classical PC is larger than the its robust counterparts. This means that the first component has been inflated by outliers.

Some other differences are evident for the second and third component. In particular, for the second component, the variable `stroke` has a very small loading with classical PCA, while the largest (in absolute value) with robust methods. In order to be able to interpret the robust components, it shall be noted that the first PC contrasts the dimensions of the vehicles and the characteristics of the engine (positive sign) with fuel efficiency (negative sign). The `price` appears to be related to the former set of variables, given it has a positive sign. The second and third component are mainly dominated by the variables `stroke`, `compression-ratio` and `height`.

Figure 3.6 shows scatterplots of the scores based on classical and RMCD-PCA. We focus on the first three main directions. A different symbol has been used for cars running on diesel. In both panels, the 97.5% tolerance ellipse has been included. As a general rule, data points that fall inside the ellipse (or not very far from it) can be classified as regular points. Many points corresponding to cars running on diesel are inside the classical tolerance ellipses and thus are not *separated* from the others. The shape of the tolerance ellipses is inflated to accommodate these outliers. On the contrary, the points corresponding to cars running on diesel fall outside the robust tolerance ellipses and are clearly grouped together. There are also other points that lie far from the fitted tolerance ellipse. These points deserve further attention: they might be outliers for unforeseen reasons (rather than because of diesel engines). A further investigation of these points may lead to discovering other groups of cars sharing a characteristic making them different from the rest of the data. Finally, we notice that the third robust component is particularly effective in identifying the group of cars running on diesel.

TABLE 3.2
Automobile data. Loadings of the first three components based on PCA, MCD-PCA and MM-PCA. The last line gives the cumulative percentage of explained (robust) variance.

	PCA			RMCD-PCA			MM-PCA		
	PC1	PC2	PC3	PC1	PC2	PC3	PC1	PC2	PC3
symboling	0.09	-0.39	-0.34	-0.07	0.26	-0.34	-0.07	0.25	-0.30
wheel-base	-0.29	0.29	0.10	0.31	-0.17	0.27	0.31	-0.17	0.25
length	-0.33	0.15	0.09	0.39	-0.15	0.14	0.40	-0.15	0.15
width	-0.32	0.09	-0.08	0.27	-0.02	0.06	0.29	-0.03	0.07
height	-0.12	0.41	0.38	0.14	-0.35	0.42	0.15	-0.34	0.42
curb-weight	-0.35	0.03	-0.08	0.29	-0.01	-0.03	0.29	-0.01	-0.04
engine-size	-0.32	-0.11	-0.24	0.29	0.07	-0.06	0.28	0.07	-0.06
bore	-0.26	-0.02	0.09	0.21	-0.15	-0.17	0.21	-0.14	-0.17
stroke	-0.05	0.07	-0.57	0.12	0.75	0.53	0.12	0.76	0.52
compression-ratio	-0.02	0.43	-0.43	-0.20	-0.34	0.41	-0.21	-0.34	0.45
horsepower	-0.29	-0.31	-0.08	0.31	0.11	-0.13	0.29	0.12	-0.13
peak-rpm	0.08	-0.38	0.24	-0.08	0.19	0.26	-0.06	0.18	0.28
city-mpg	0.30	0.26	-0.16	-0.28	-0.06	0.13	-0.28	-0.06	0.13
highway-mpg	0.32	0.21	-0.15	-0.29	-0.04	0.14	-0.29	-0.04	0.13
price	-0.32	-0.10	-0.14	0.34	-0.04	-0.06	0.33	-0.03	-0.04
%	50.59	67.64	76.47	42.62	62.13	73.29	43.88	63.61	74.50

3.4 R Illustration (Robust scoreplot of automobile data) *The robust scoreplot based on RMCD in the first bottom panel of Figure 3.6 was obtained with the following code:*

```
> plot(mcd.acp@scores[Automobile[,2]=='gas',2],
+ mcd.acp@scores[Automobile[,2]=='gas',2],
+ xlim=range(mcd.acp@scores[,1]),
+ ylim=range(mcd.acp@scores[,2]),
+ ylab='PC1', xlab='PC2')
> points(mcd.acp@scores[Automobile[,2]=='diesel',1],
+ mcd.acp@scores[Automobile[,2]=='diesel',2],pch='+')
> abline(v=0,h=0)
> lines(ellipse(x=diag(mcd.acp@eigenvalues[1:2]),
+ level=0.975))
```

The fact that classical PCA is unreliable in presence of outliers, and that there is masking during outlier identification, is stressed further when building outlier maps. Outlier maps for the Automobile data are shown in Figure 3.7. The left panel corresponds to classical PCA: the cars running on diesel are all masked. On the contrary, in the outlier map based on RMCD (right panel), cars running on diesel are all flagged as bad leverage points and well separated from the others.

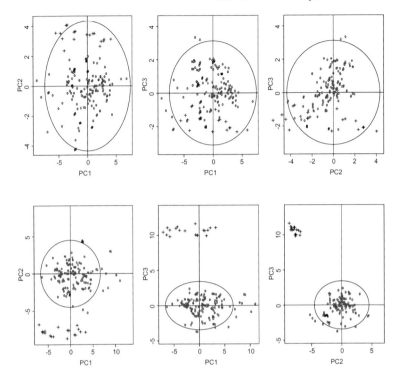

FIGURE 3.6
Automobile data. Scoreplot and 97.5% tolerance ellipse based on PCA (top)
and RMCD-PCA (bottom) for the first three components.

Other suspicious orthogonal outliers are also visible, and easily explained
after a closer look at the data at hand. Namely, observations 120, 121 and
122 are the only cars with rear engine. These three are of Porsche brand.
The large orthogonal outlier with observation number 47 is the only car with
twelve cylinders and the only Jaguar in the sample.

As well as in multivariate location and covariance estimation, robust PCA
can be enriched by *monitoring* the quantities of interest as breakdown or
efficiency values change over a grid of values. Here, we monitor both the score
and orthogonal distances based on RMCD in order to detect the minimum
BP beyond which the robust technique gives results similar to classical PCA.
As we can see in Figure 3.8, the *bad leverage* nature of the cars running on
diesel is already evident for a BP larger than 0.11.

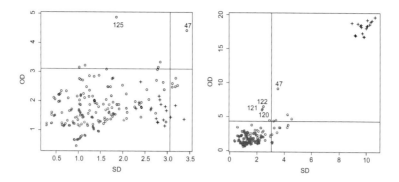

FIGURE 3.7
Automobile data. Outlier map based on classical and RMCD-PCA ($q = 3$).

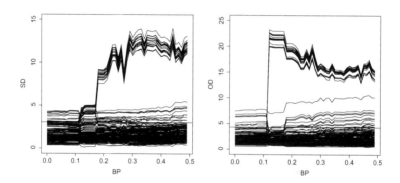

FIGURE 3.8
Automobile data. Monitoring score distances (left) and orthogonal distances (right) by varying the BP of the RMCD. The horizontal dashed line gives the corresponding cut-off values.

3.7.2   Octane data

The Octane dataset (Hubert *et al.*, 2005) is made of $n = 39$ gasoline samples, identified by their octane numbers. For each gasoline sample, the near-infrared absorbance spectra over $p = 226$ wavelengths have been measured, hence $p >> n$. Six samples $(25, 25, 36, 37, 38, 39)$ contain added alcohol and they are outliers. As before, we keep them in the sample for illustration and confirmatory analysis. The data are plotted in Figure 3.9: the six outlying spectra clearly deviate from the others, and they are all characterized by larger values on the right end of the spectrum (after the 150th wavelength).

We compare here the results of PCA, ROBPCA, robust PP and SPCA. ROBPCA is based on $\varepsilon = 0.25$ and PP is based on the MAD. The number

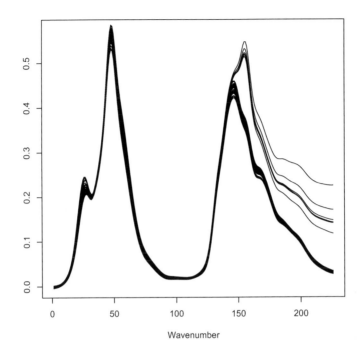

FIGURE 3.9
Octane data. Observed spectra for $n = 39$ samples.

of selected components is $q = 2$, according to Step 3 of ROBPCA. The first two loading vectors from classical PCA are displayed in Figure 3.10, whereas those from the robust methods are shown in Figure 3.11.

It can clearly be seen that the first classical component is dominated by outliers: the largest loadings correspond to the largest wavelenghts of the spectra, where the anomalies occur. Outliers, as it often happens, have also inflated the variance of the first principal component. This is shown in Table 3.3.

Robust procedures, on the other hand, manage the outliers quite well in this example. The loadings computed through the three robust methods are quite similar, and so it is the explained variability.

The score plots in Figure 3.12 highlight that points projected through classical PCA are still correlated and only the influence of the six contaminated points (marked as +) leads to the classical correlation coefficient equal to zero. Recall that PCA scores are uncorrelated by construction. Furthermore, the mean of the bulk of the data is far from the origin, although the scores have zero mean. These unpleasant features disappear when using the robust methods.

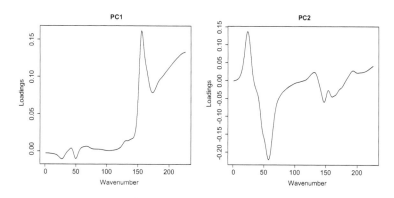

FIGURE 3.10
Octane data. Loadings of the first two components based on classical PCA.

TABLE 3.3
Octane data. Standard deviation of the first two components based on PCA, ROBPCA, PP and SPCA.

	PC 1	PC 2
PCA	0.365	0.095
ROBPCA	0.134	0.052
MAD-PP	0.151	0.062
Q_n-PP	0.157	0.066
SPCA	0.146	0.040

The six outlying octane samples are clearly flagged by the outlier maps in Figure 3.13 when these are based on robust estimates. Outliers are all flagged as bad leverage points. Classical PCA, instead, leads to a huge masking and only one gasoline sample is detected as anomalous.

The nature and influence of the six outliers can be explained further by looking at Figure 3.14, which shows the spectra approximated by PCA and ROBPCA based on $q = 2$. The robustly approximated spectra for the outlying samples are all very similar to the general pattern. We notice a large difference between the observed and fitted spectra for the six outlying samples. This is located at large wavelengths, and is in agreement with the marking of them as bad leverage points.

3.5 R Illustration (Robust PCA of octane data) *ROBPCA, PP, SPCA were obtained with the following code. More details on the options available are given in Appendix A.1.*

```
> rob.H<-PcaHubert(X, k=2,alpha=0.75,scale=FALSE)
```

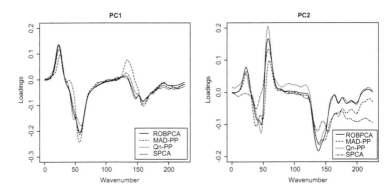

FIGURE 3.11
Octane data. Loadings of the first two components based on ROBPCA, MAD-PP, Q_n-PP and SPCA.

```
> rob.Loc<-PcaLocantore(X,k=2)
> rob.pp<-PCAgrid(X,k=2,method='mad')
```

The outlier maps in Figure 3.13 based on ROBPCA and SPCA can be obtained through the use of the `plot` *function on the output of the functions above. The outlier map based on PP was obtained with the following code*

```
> out<-PCdiagplot(X,rob.pp,plot=F,raw=T,
+ ksel=2,crit=0.975)
> plot(out$SDist[,1],out$ODist[,1],
+ ylab='OD', xlab='SD',main='PP')
> abline(h=out$critOD,v=out$critSD)
```

3.7.3 Video surveillance data

These data have been analyzed by De La Torre and Black (2003), who formulated PCA as a least squares problem and then replaced ordinary regression with Huber type robust regression and by Mateos and Giannakis (2010), who modified the PCA problem by directly accounting for the number of outliers. Here, background learning is accomplished by ROBPCA and PP based on the MAD.

We only consider the first $n = 100$ video frames of resolution 120×160 pixels. This resolution maps into $p = 19200$ variables. Some of the images captured people in various locations. Some people pass through the visual of the camera quickly, some other remain relatively still over multiple frames. Then, areas of pixels corresponding to persons can be considered outliers and identified through outlier detection. The objective is the extraction of a low-rank stationary background model for use in person detection and tracking.

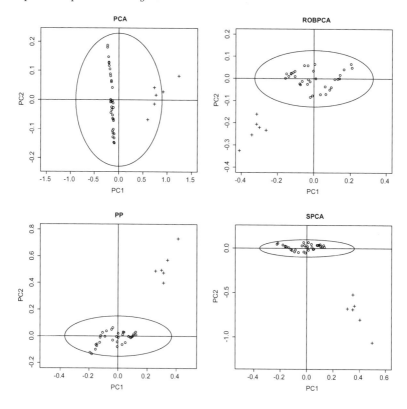

FIGURE 3.12

Octane data. Score plot based on PCA, ROBPCA, MAD-PP and SPCA, including robust 97.5% tolerance ellipses.

We obtain the desired low dimensional reconstruction of the data by the first $q = 5$ robust loading vectors. The results are shown in Figure 3.15 for a selection of frames. The robust result is also compared to the ordinary PCA performed by SVD and based on the first $q = 5$ dominant eigenvectors, as well. The first column shows the original frames, the second the classical PCA image reconstruction, the third and fifth columns give the ROBPCA and PP image reconstruction, respectively, the fourth and sixth columns display the corresponding residual matrices. It can be seen that classical PCA fails in separating the background from people: the presence of subjects results in *ghostly* artifacts that badly influence the estimate of the background. On the other hand, ROBPCA and PP are able to capture the background and separate it from the individuals. It shall be noted that if we increase q, classical PCA does a better job in recovering the background, but at the price of a less parsimonious representation and still with a worse performance in outlier identification than robust methods based on $q = 5$.

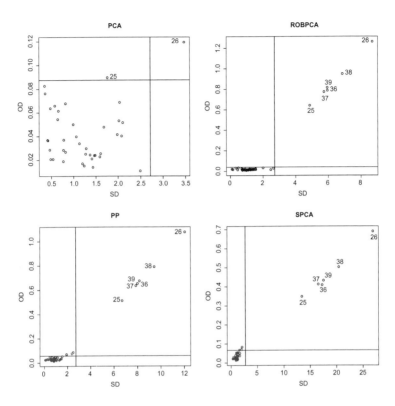

FIGURE 3.13
Octane data. Outlier map based on PCA, ROBPCA, MAD-PP and SPCA.

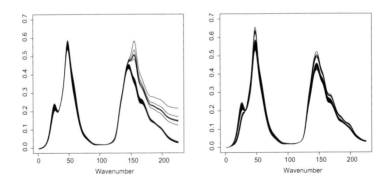

FIGURE 3.14
Octane data. Spectra from PCA (left) and ROBPCA (right) based on the first two components.

Original	PCA	ROBPCA		PP	

FIGURE 3.15

Video surveillance data. First column: original frame. Second column: PCA. Third column: ROBPCA. Fourth column: ROBPCA residual frame, Fifth column: PP. Sixth column: PP residual frame.

4

Sparse Robust PCA

The interpretation and labeling of the principal components, regardless of whether they arise from robust or classical PCA, is often difficult. A source of difficulty is the fact that *all* the loadings are nonzero, since each PC is a linear combination of *all* the original variables. The common approach to interpretation of principal components is that of ignoring features associated with loadings that are small in absolute value. This practice can be shown to be misleading and sub-optimal (Cadima and Jolliffe, 1995).

These facts have motivated the introduction of different methods leading to loading matrices where some entries are exactly zero. These loading matrices are still optimal with respect to a certain objective function which somehow includes the requirement of zero loadings (e.g., through penalties).

Such a loading matrix is called *sparse* and leads to a sparse PCA. Sparse PCA is conceived to lead to components that are linear combination of only a small subset of the original variables. Those variables giving a null contribution on the first q components could be discarded. Therefore sparse PCA also possibly performs variable selection and is particularly useful in high dimensions, when many variables might be redundant. Sparse PCA has a wide array of applications, from gene selection in genomics, to finance, machine learning, engineering and computer vision. Sparse matrices are better stored and handled in computers, therefore sparse PCA usually leads to better compression of the data matrices with respect to non-sparse methods.

These ideas are not recent and have previously motivated the development of *rotation* procedures for principal components. Rotation is used to decrease small loadings. Still, rotation procedures do not give exactly zero loadings. Formal methods for sparse PCA are always to be preferred to two step procedures where rotation of the loadings follows classical or robust components extraction.

Generally speaking, sparse methods are obtained by suitably modifying maximization criteria leading to classical PCA. This is done by introducing constraints on the loadings (e.g., on their L_1 or even L_0 norm). Sensitivity of unconstrained classical PCA to the occurrence of outliers is inherited by its sparse counterpart. This motivated the development of robust methods which would also lead to sparse loading matrices. As before, robust methods can be based on robust estimates of the covariance matrix, or on projection pursuit.

4.1   Basic concepts and sPCA

The main drawback of dimensionality reduction through PCA is that each principal component is a linear combination of *all p* original variables. This may drastically complicate the interpretation and labeling of the components. Actually, if each principal component were a linear combination of only a small number of the original variables, the subjective interpretation step would be much easier. An additional useful feature could be that each original variable would be used to build only one or very few principal components. A principal component whose loading vector includes some zeros is called *sparse*. Simpler interpretation is not the only advantage of *sparseness* of the loadings. In fact, variables receiving a zero loading in all the q selected components can be discarded. Moreover, sparse matrices are more easily stored and handled.

A common and often intrinsic approach to build sparse principal components is given by setting to zero all loadings under a fixed threshold: this is called *simple thresholding* of the loadings. Simple thresholding is anyway potentially misleading and does not produce results which are optimal under any criterion. Furthermore, simple thresholding leads to correlated and collinear components. For a full account on the topic, see Cadima and Jolliffe (1995).

Sparse principal components should be obtained from more formal methods. One of the first approaches is the SCoTLASS (Jolliffe *et al.*, 2003): the variance of the components is maximized under the constraint of an upper bound on the sum of the absolute value of the loadings. The methodology makes use of non-convex constrained optimization. An additional constraint can be used to impose orthogonality of the loading vectors.

Formally, the SCoTLASS criterion to find the jth sparse loading vector is

$$\nu_j = \operatorname{argmax}_{||a||=1, a \perp \nu_1 \perp \dots \perp \nu_{j-1}} a^{\mathrm{T}} S a - \lambda_{1,j} ||a||_1, \tag{4.1}$$

where $||a||_1$ denotes the L_1 norm of the vector a. Here we introduce a penalty parameter $\lambda_{1,j}$, which should not be mistaken for the eigenvalues of the covariance matrix $\lambda_1, \dots, \lambda_p$. The penalty parameter should be chosen by the user and it is fixed *a priori*. The larger $\lambda_{1,j}$, the more sparse is the resulting component.

Another procedure to obtain sparse PCA is given by the sPCA method described in Zou *et al.* (2006), who frame the PCA problem as ridge regression and propose an algorithm based on a modified elastic net (Zou and Hastie, 2005). Elastic net combines L_2 and L_1 penalization terms: the L_1 penalty leads to sparse approximations of the ordinary components, (see also Leng and Wang (2009) and Guo *et al.* (2010) for further improvements and Witten *et al.* (2009) for a more general approach). The computational cost of sPCA is much lower than SCoTLASS and often sPCA leads to the same amount of variability in the components with more sparsity. In the following, sPCA is presented more in details.

Let $A = [\alpha_1, \dots, \alpha_q]$ and $B = [\beta_1, \dots, \beta_q]$ be $p \times q$ matrices. Without loss of generality, as usual, we assume the column means of X are equal to zero. It is shown in Zou *et al.* (2006) that the first q principal components of X are the solution to

$$(\hat{A}, \hat{B}) = \arg\min_{A,B} \sum_{i=1}^{n} ||x_i - x_i AB^{\mathrm{T}}||^2 + \lambda_2 \sum_{j=1}^{q} ||\beta_j||^2 \qquad (4.2)$$

subject to $A^{\mathrm{T}} A = I_q$, where I_q is the identity matrix of dimension q and λ_2 is a fixed penalty parameter. The elements of $\hat{B}$ which satisfy the criterion are proportional to the loadings. A sparse approximation of the loadings is obtained by adding an L_1 (LASSO) penalty into (4.2), leading to the criterion

$$(\hat{A}, \hat{B}) = \arg\min_{A,B} \sum_{i=1}^{n} ||x_i - x_i AB^{\mathrm{T}}||^2 + \lambda_2 \sum_{j=1}^{q} ||\beta_j||^2 + \sum_{j=1}^{q} \lambda_{1,j} ||\beta_j||_1 \ . \quad (4.3)$$

As anticipated, in (4.3) we have $q+1$ penalty parameters, $\lambda_{1,j}$ for $j = 1, \ldots, q$ and λ_2. The first q penalty parameters control the degree of sparsity of each loading vector.

The optimization in (4.3) is then simplified by noting that

$$
\begin{aligned}
\sum_{i=1}^{n} ||x_i - x_i AB^{\mathrm{T}}||^2 &= ||X - XBA^T||^2 \\
&= ||XA_\perp||^2 + ||XA - XB||^2 \\
&= ||XA_\perp||^2 + \sum_{j=1}^{q} ||X\alpha_j - X\beta_j||^2
\end{aligned}
$$

where $A_\perp$ is any orthonormal matrix such that $[A; A_\perp]$ is $p \times p$ and orthonormal. Therefore, the sPCA criterion (4.3) only depends on the original data matrix via the sample covariance matrix $S = n^{-1}X^{\mathrm{T}}X$, since

$$||X\alpha_j - X\beta_j||^2 = (\alpha_j - \beta_j)^{\mathrm{T}} X^T X (\alpha_j - \beta_j), \ j = 1, 2, \ldots, q \ . \qquad (4.4)$$

In conclusion, the elastic net regression problem (4.3) is solved by an alternating least squares algorithm whose general iteration is given in Algorithm 4.1. An implementation is available in the R library **elasticnet**, that contains the **spca** function to perform sPCA.

The sparse PCs obtained from sPCA are neither orthogonal nor uncorrelated. Only the loadings resulting from ordinary PCA can be orthogonal and simultaneously yield uncorrelated components (Jolliffe, 2005). One of the limitations of having correlated components is that the variance accounted for by the first q components does not correspond to the sum of their variances (as with classical PCA). Actually, sparsity of the loadings induces additional information loss: the variability of the sparse component is lower than that of the unconstrained component. Zou *et al.* (2006) propose to measure the

---

Algorithm 4.1 sPCA

Update B
for $j = 1, \ldots, q$ **do**
$$\hat{\beta}_j = \operatorname{argmin}_\beta (\alpha_j - \beta)^{\mathrm{T}} S(\alpha_j - \beta) + \lambda_2 ||\beta||^2 + \lambda_{1,j} ||\beta||_1$$
end for
Update A
Solve $SB = UDV^{\mathrm{T}}$.
Let $\hat{A} = UV^{\mathrm{T}}$

---

information contained in the first q sparse principal components using the *adjusted variance*. This is defined as follows: let r_j be the vector of residuals of the linear regression that has the jth PC as response and the previous ones as predictors. The total variance explained by the first q PCs is $\sum_{j=1}^{q} ||r_j||^2$.

4.1 Example (sPCA of spam data) *Let us consider the data on spam detection. The loadings of the first classical non-sparse principal component are shown in Figure 4.1. The first PC accounts for 17.7% of the total variability. It is apparent that many loadings are close to zero. Therefore, sPCA may succeed in setting some of those loadings exactly equal to zero.*

Figure 4.2 shows the percentage of explained variance as the degree of sparsity of the first component ranges from full sparseness (the component is uniquely determined by one variable) to unconstrained PCA. One can conclude that if the first component is obtained as a linear combination of only three variables there might be a negligible loss of information. With three variables the percentage of explained variance drops only slightly, from 17.7% to 17.2%.

4.2 R Illustration (sPCA of spam data) *Figure 4.2 was obtained with the following code. See section A.3 for further details on the function* spca.

```
> library(elasticnet)
> sparsity<-1:20
> pev<-NULL
> for(i in 1:20)
+       {out<-spca(spam20, K=1,type='predict',
+       para=sparsity[i],sparse='varnum')
+       pev[i]<-out$pev}
> plot(sparsity, pev, type='b',
+ xlab='Number of nonnull loadings',
+ ylab='Variance Explained')
```

We conclude mentioning that there are now many other approaches to sparse PCA, among which we also cite the contributions from Farcomeni

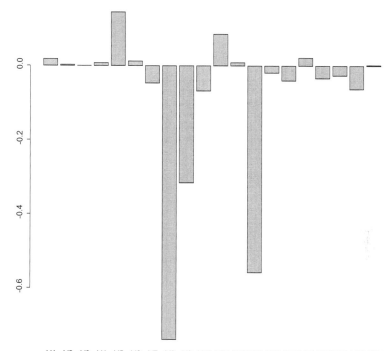

V1 V2 V3 V4 V5 V6 V7 V8 V9 V10 V11 V12 V13 V14 V15 V16 V17 V18 V19 V20

FIGURE 4.1
Spam data. Loadings of the first classical non-sparse component.

(2009a), d'Aspremont *et al.* (2008, 2014), Shen and Huang (2008), Qi *et al.* (2013) and Merola (2014). The approach of Farcomeni (2009a), for instance, is based on an efficient bounding algorithm for exactly solving the L_0 penalized sparse PCA problem. Explicit constraints can be used to choose the number of nonzero loadings for each component and to impose orthogonality *or* zero correlation of the principal components.

4.2 Robust sPCA

The optimization problems in (4.1) and (4.3) are clearly non-robust with respect to the occurrence of anomalous values in the data because of the sensitivity of the sample covariance matrix S. An obvious way for combining sparseness and robustness would be to replace S by a robust estimate $\hat{\Sigma}$. A

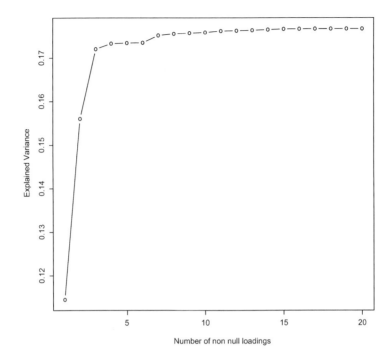

FIGURE 4.2
Spam data. Percentage of variance explained by the first component for varying number of non null loadings.

plug-in robust counterpart of the SCoTLASS criterion (4.1) has been discussed in Croux *et al.* (2013).

In this section we develop a plug-in robust sPCA algorithm. The initial values for α_j, $j = 1, 2, \ldots, q$ to initialize Algorithm 4.1 can correspond to the leading eigenvectors of $\hat{\Sigma}$. Minimization of the objective function

$$\sum_{j=1}^{q} (\alpha_j - \beta_j)^{\mathrm{T}} \hat{\Sigma} (\alpha_j - \beta_j) + \lambda_2 ||\beta||^2 + \lambda_{1,j} ||\beta||_1$$

yields sparse and robust components that approximate the dominant eigenvectors of $\hat{\Sigma}$, where $\hat{\Sigma}$ is a robust estimate of the covariance matrix. Algorithm 4.1 is feasible as long as we are able to compute a robust estimate $\hat{\Sigma}$.

The scores from robust sPCA are mildly correlated, as well as in the non-robust scenario. As a consequence, the amount of explained variance should be estimated similarly to Section 4.1. Our proposal is to use a robust regression scheme, such as Least Trimmed Squares regression (Rousseeuw, 1984;

Rousseeuw and Leroy, 1987; Rousseeuw and Van Driessen, 1999; Pison *et al.*, 2002) or MM regression (Rousseeuw and Yohai, 1984; Salibian-Barrera and Yohai, 2006; Maronna *et al.*, 2006). The total variance explained by the first q PCs is $\sum_{j=1}^{q} \hat{\sigma}^2(r_j)$, according to the robust measure of scale chosen.

A by-product of using (robust) regression to estimate the variance is that *new* scores are computed. These are the residuals r_j. They contain slightly less information than the (robust) sparse scores, but they have approximately zero correlation. An implementation of this strategy is straightforward as software routines for robust regression are already available in the R package `robustbase` (but also in package `MASS`). LTS regression, can be performed using function `ltsReg` and MM regression with function `lmrob`.

4.3 Choice of the degree of sparsity

The choice of the penalty parameter $\lambda_{1,j}$, $j = 1, \ldots, q$, controlling the degree of sparsity of the jth component is an open issue. More sparsity corresponds to larger parameters, but it is not possible to know in advance how many zeros will be obtained for a given $\lambda_{1,j}$. Some guidelines for choosing $\lambda_{1,j}$ in the non-robust case can be found in Leng and Wang (2009), Farcomeni (2009a) and Guo *et al.* (2010).

In practice, finding q different optimal sparsity parameters can be computationally expensive. A simple shortcut is to set $\lambda_{1,j} = \lambda, \forall\, j$. Here, we consider the robust criterion proposed by Croux *et al.* (2013). Let $\tilde{V}_q$ and $\hat{V}_q$ be the loading matrices from sparse and unconstrained PCA containing the first q PC directions, respectively. Denote by $\tilde{R} = X - X\tilde{V}\tilde{V}^T$ and $\hat{R} = X - X\hat{V}\hat{V}^T$ the corresponding residual matrices, whose j^{th} columns are $\tilde{r}_j = (\tilde{r}_{1j}, \tilde{r}_{2j}, \ldots, \tilde{r}_{nj})^{\mathrm{T}}$ and $\hat{r}_j = (\hat{r}_{1j}, \hat{r}_{2j}, \ldots, \hat{r}_{nj})^{\mathrm{T}}$. A robust criterion for selection is given by

$$Q_{CFF}(\lambda) = \frac{\sum_{j=1}^{q} \hat{\sigma}^2(\tilde{r}_{1j}, \tilde{r}_{2j}, \ldots, \tilde{r}_{nj})}{\sum_{j=1}^{p} \hat{\sigma}^2(\hat{r}_{1j}, \hat{r}_{2j}, \ldots, \hat{r}_{nj})} + m(\lambda)\frac{\log n}{n}, \qquad (4.5)$$

where $m(\lambda)$ denotes the number of nonzero loadings in $\tilde{V}_q$.

It is worth noting that a selection criterion can be obtained based directly on choosing the degree of sparsity in each component. This feature is particularly attractive in the case of sPCA and sRPCA, since the R function `spca` also allows to fix the cardinality (number of non null loadings) of each component as an alternative to fix the penalty parameters. Such an approach has been proposed by Farcomeni (2009a) who suggested to select the number of nonzero loadings m for the jth sparse component by maximizing

$$Q_F = J(S, \hat{\beta}_j) - \frac{\bar{\sigma}^2}{p}\frac{\log m}{j+1}$$

with respect to m. In (4.3) $J(\cdot)$ denotes the variance (for the first component) or the adjusted variance (for the other components) and $\bar{\sigma}^2 = \frac{\sum_{j=1}^{q} \hat{\Sigma}_{jj}}{q}$.

4.4   Sparse projection pursuit

Similarly to the methods based on a robust estimate of the covariance matrix described in the previous sections, the PP approach to PCA discussed in Section 3.3 can be modified to lead to sparse dimension reduction by adding an L_1 penalty in (3.1) and (3.2), yielding

$$\nu_{\hat{\sigma};1} = \text{argmax}_{||a||=1}\hat{\sigma}^2(a^{\text{T}}x_1, a^{\text{T}}x_2, \ldots, a^{\text{T}}x_n) - \lambda_{1,1}||a||_1 \qquad (4.6)$$

for the first component and for $j = 2, \ldots, q$

$$\nu_{\hat{\sigma};j} = \text{argmax}_{||a||=1}\hat{\sigma}^2(a^{\text{T}}x_1, a^{\text{T}}x_2, \ldots, a^{\text{T}}x_n)) - \lambda_{1,j}||a||_1, \qquad (4.7)$$

under the additional constraint that $\nu_{\hat{\sigma};1} \perp \ldots \perp \nu_{\hat{\sigma};j}$. For instance, $\hat{\sigma}^2$ can be the squared MAD or the squared Q_n. This methodology (sPP) has been proposed in Croux *et al.* (2013), which to our knowledge is the first contribution towards a dimension reduction method both robust and sparse.

As well as for unconstrained PCA, the sPP technique does not need computing an estimate of covariance. Sparse components are obtained sequentially, thus substantially reducing the computational burden when $q << p$. An efficient and fast algorithm for solving the robust optimization problems in (4.6) and (4.7) is described in Croux *et al.* (2013).

The PP approach to sparse PCA is very effective when p is large and/or $p > n$. The sPP methodology is available in the R packages pcaPP and rrcovHD, through the functions sPCAgrid and SPcaGrid, respectively.

Two methods can be implemented to select the optimal tuning parameter: one is based on (4.5), and is implemented in function opt.BIC; the other is based on a trade-off curve proposed in Croux *et al.* (2013).

The trade-off curve plots the percentage of explained (robust) variance as a function of λ or, equivalently, of the number of zero loadings. The robust explained variance is measured as

$$EV_q = \frac{\sum_{j=1}^{q} \hat{\sigma}^2(Y_j)}{\sum_{j=1}^{p} \hat{\sigma}^2(X_j)}. \qquad (4.8)$$

When $\hat{\sigma}^2$ coincides with the sample variance, (4.8) corresponds to the ratio of the sum of the q largest eigenvalues and the sum of all the eigenvalues of the sample covariance matrix, as usual. The trade-off curve is computationally efficient and allows us to estimate different tuning parameters for

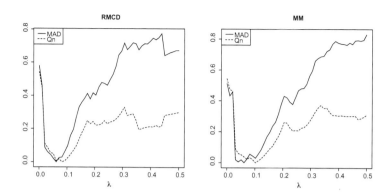

FIGURE 4.3
Automobile data. Selection of the penalty parameter.

each component. As usual, an elbow indicates the optimal value of the tuning parameter. The criterion is available through function `opt.TPO` (Trade-off Product Optimization). A heuristic criterion selects the tuning parameters by maximizing the explained variance multiplied by the number of zero loadings of a particular component.

4.5 Examples

Here we illustrate the sRPCA and sPP methods, along with the related diagnostic plots, through some real data examples.

4.5.1 Automobile data

Let us consider the data from the Automobile Imports Database. In order to obtain a sparse representation of the robust loadings, we apply the sRPCA algorithm based on both the 25% breakdown point RMCD and the 95% location-efficient MM estimate of the covariance matrix, considered in Section 3.7.1. The penalty parameter has been selected by using the criterion (4.5) based on the Q_n measure of scale, rather than on the MAD. The reason is that the use of the former leads to a larger sparsity parameter (hence larger sparseness of the loadings), with a negligible loss of information.

The criterion (4.5) is shown in Figure 4.3 over a grid of 50 values: the minimum based on the Q_n occurs at $\lambda = 0.082$ for the RMCD and $\lambda = 0.102$ for MM estimation.

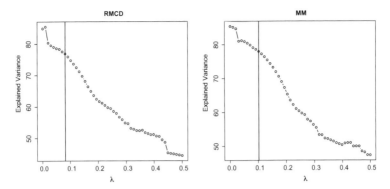

FIGURE 4.4
Automobile data. Trade off curve for sRPCA based on RMCD (left) and MM (right). The dashed line represents the penalty parameter selected by optimization of Q_{CFF}.

The resulting sparse loadings are given in Table 4.1. The results based on the use of the RMCD and the MM are very similar. The occurrence of zero loadings clearly enhances the interpretation of the components as compared to those given in Table 3.2. This is especially true for what concerns the second and third component: the former only contrasts four vehicle characteristics including `compression ratio` with the actuarial variable `symboling`, the latter is dominated by the variable `stroke`, that does not appear in the other components.

The last line of Table 4.1 reports the percentage of explained adjusted variance after orthogonalization has been performed by LTS regression (with 0.25 rate of trimming). The total variance has been obtained by means of the univariate LTS estimator of scale (that corresponds to the univariate MCD).

Figure 4.4 shows the proportion of variance explained by the $q = 3$ components as a function of the penalty parameter. It can be seen that percentage of explained variance drops slightly at the selected penalty. On the other hand, at the selected penalty we obtain a slightly large number of null loadings, which are three for the first component, ten for the second and thirteen for the third. We conclude that the trade off between sparsity and information loss is well acceptable.

The scores are shown in Figure 4.5 for the case of MM-sRPCA. In simple words, a covariance estimate is computed based on MM, then sPCA is used to compute sparse scores (left panel). The resulting scores are used in a LTS regression framework to compute residuals. These residuals are based on sparse scores, and are approximately uncorrelated as illustrated in the right panel. The orthogonalization effect is particularly evident for the third PC.

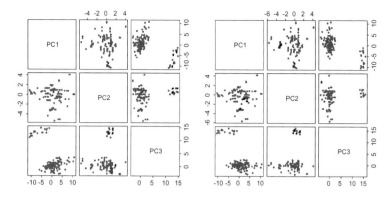

FIGURE 4.5
Automobile data. Scores and orthogonal scores (residuals from robust regression) based on MM-sRPCA and LTS regression.

After we have introduced the sparsity constraint in the robust PCA, it is natural to wonder if its diagnostic power has been weakened somehow. To this end, we first compute score and orthogonal distances from the sparse and robust fit and then represent them in an outlier map. It can be seen from

TABLE 4.1
Automobile data. Loadings of the first three components for sRPCA based on RMCD and MM. The last line gives the cumulative percentage of variance explained by sparse robust PCA.

	RMCD			MM		
	PC1	PC2	PC3	PC1	PC2	PC3
symboling	0	-0.42	0	0	-0.38	0
wheel-base	0.17	0.43	0	0.18	0.43	0
length	0.43	0.23	0	0.45	0.25	0
width	0.22	0	0	0.24	0	0
height	0	0.59	0	0	0.59	0
curb-weight	0.15	0	0	0.15	0	0
engine-size	0.40	0	0	0.39	0	0
bore	0.17	0	0	0.16	0	0
stroke	0	0	0.98	0	0	0.97
compression-ratio	-0.33	0.49	0	-0.35	0.52	0
horsepower	0.32	0	0	0.29	0	0
peak-rpm	-0.07	0	0.22	-0.05	0	0.23
city-mpg	-0.20	0	0	-0.19	0	0
highway-mpg	-0.36	0	0	-0.38	0	0
price	0.37	0	0	0.35	0	0
% variance	55.78	13.49	7.56	56.98	13.44	7.55

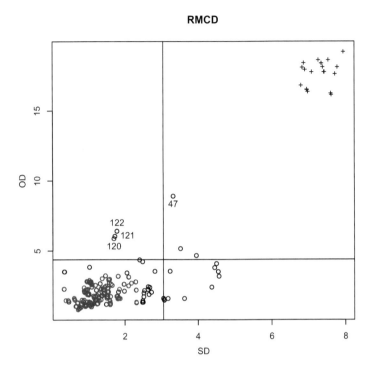

FIGURE 4.6
Automobile data. Outlier map based on RMCD-sRPCA.

Figure 4.6 that sparsity constraints do not affect the diagnostic tool. The outlier map is very similar to that obtained without sparsity constraints, but the sparse solution is more interpretable. The outlier map based on sparse robust PCA still correctly flags the group of cars running on diesel.

4.3 R Illustration (sRPCA of automobile data) *The loadings of RMCD and MM based sRPCA were obtained with the following code. In the following* est *is the robust estimate of covariance and* lambda.opt *is the optimal penalty parameter. The* type *option has been set to* Gram *since the* x *argument is a covariance matrix.*

```
> library(elasticnet)
> out <- spca(x=est, K=3, para=rep(lambda.opt,3),
+ type='Gram',sparse='penalty')
```

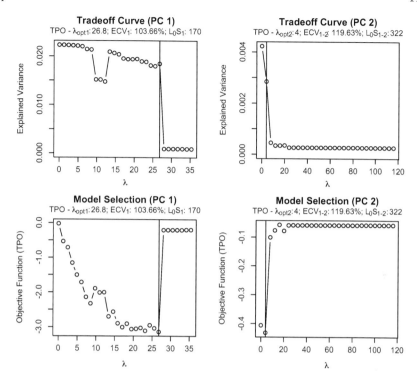

FIGURE 4.7
Octane data. Trade-off curve and TPO for sPP based on MAD.

4.5.2 Octane data

We now work with the Octane data. Given the large number of measurements ($p = 226$), a sparse representation of the loadings could be useful for identifying relevant spectral ranges. We make use of sPP and choose both the MAD and the Q_n as robust projection indexes.

Furthermore, we select a different tuning parameter for each sparse component. The tuning parameters $\lambda_{1,j}, j = 1, 2$ have been selected according to the trade-off curve. The TPO selection procedure is shown in Figures 4.7 and 4.8: the penalty parameter chosen for each component is located right before the drop of the trade off curve (top panels), which makes the TPO (with negative sign) attain its minimum (bottom panels).

The sparse structures of the loadings resulting from the use of the MAD and the Q_n are displayed in Figures 4.9 and 4.10, respectively. The number of non null loadings of the first and second sparse components based on the MAD is 56 and 74, respectively, whereas Q_n leads to a less sparse representation with 93 and 120 non null loadings.

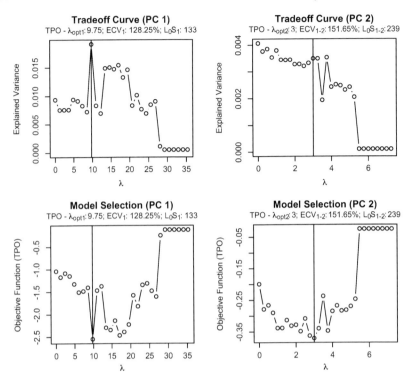

FIGURE 4.8
Octane data. Trade-off curve and TPO for sPP based on Q_n.

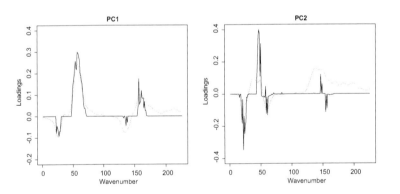

FIGURE 4.9
Octane data. Loadings of the first two components for sPP based on MAD.
The gray lines show the non-sparse results.

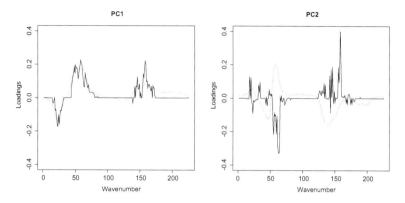

FIGURE 4.10
Octane data. Loadings of the first two components for sPP based on Q_n. The gray lines show the non-sparse results.

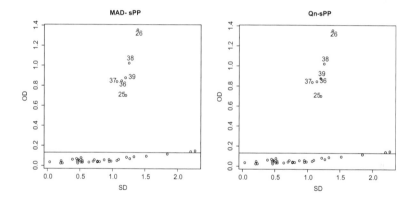

FIGURE 4.11
Octane data. Outlier map based on MAD-sPP and Q_n-sPP.

The sparse loadings suggest that there are some wavelengths that are clearly more relevant in explaining the total variability, and many other that may be discarded in future studies.

The outlier maps given in Figure 4.11 show that the sparse methods identify data anomalies, and flag them as orthogonal outliers. It should be noted that the score distances are somewhat smaller than those resulting from the unconstrained robust methodology.

4.4 R Illustration (sPP of octane data) *The penalty parameters characterizing sPP based on MAD and Q_n were obtained with the following code. The data are stored in the matrix* X.

```
> library(pcaPP)
> obj.mad <- opt.TPO (X, k.max = 2, method='mad')
> obj.qn <- opt.TPO (X, k.max = 2, method='qn')
```

Figure 4.7 was obtained with the following code:

```
> par(mfrow=c(2,2))
> for (i in 1:2)  plot (obj.mad, k = i, f.x = "lambda")
> for (i in 1:2)  objplot (obj.mad, k = i)
```

In a similar fashion one can obtain Figure 4.8. The output from opt.TPO *already gives the estimated loadings but the sparse components can be also obtained as follows*

```
> spp <- sPCAgrid(X, k=2, lambda=obj.mad$pc$lambda,
+ method='mad')
```

The argument lambda *can be used to specify the desired tuning parameter.*
    The left panel of Figure 4.11 was obtained with the following code:

```
> out.mad.sp<-PCdiagplot(X,out.mad,plot=F,
+ raw=T,ksel=2,crit=0.975 )
> plot(out.mad.sp$SDist[,1],out.mad.sp$ODist[,1],
+ ylab='OD', xlab='SD',main='MAD-sPP')
> abline(h=out.mad.sp$critOD,v=out.mad.sp$critSD)
```

The same code can be adapted to display the outlier map resulting from the use of the Q_n estimator.
    See Section A.3 of the Appendix for further details on the functions used above.

5

Canonical Correlation Analysis

This chapter deals with Canonical Correlation Analysis (CCA), a dimension reduction methodology dating back at least to Hotelling (1936). CCA is used to assess the association between two sets of variables.

The main idea is that of summarizing the dependence between two sets of variables through the maximal correlation that can be found between two linear combinations, one made of variables from the first set, and another made of variables from the second set. The classical method (see e.g., Johnson and Wichern (2002)) relies on the sample covariance/correlation matrix, therefore being clearly not robust with respect to contamination in the data. In parallel with classical PCA, even a single outlier can completely break down the estimate of the covariance/correlation matrix, hence leading to results that are entirely unreliable. The robust alternatives that will be illustrated in the following sections are based on some of the ideas discussed in the previous chapters, such as robust covariance estimation methods and projection pursuit. The reader is also referred to Dehon *et al.* (2000), Oliveira *et al.* (2004) and Branco *et al.* (2005) for further developments on robust CCA.

5.1 Classical canonical correlation analysis

Canonical correlation analysis is a tool to investigate the relationship between two sets of variables. These associations are summarized by measuring the *correlation* between a pair of linear combinations of that variables, one for each set.

Let $X_1 = (X_{11}, X_{12}, \ldots, X_{1p_1})^{\mathrm{T}}$ and $X_2 = (X_{21}, X_{22}, \ldots, X_{2p_2})^{\mathrm{T}}$ denote the two set of measurements, whose dimension is p_1 and p_2, respectively. If we let $X = [X_1, \ X_2]$, the sample covariance matrix of X can be partitioned as follows:

$$S = \begin{pmatrix} S_{11} & S_{12} \\ S_{21} & S_{22} \end{pmatrix},$$

where S_{11} is the covariance matrix of X_1, S_{22} is the covariance matrix of X_2, S_{12} contains the cross covariances, and $S_{21} = S_{12}^{\mathrm{T}}$. The matrix S_{12} summarizes the association between the two sets of variables with $p_1 p_2$ elements. CCA aims at measuring the association summarized by S_{12} with a reduced summary.

This is performed to aid interpretation. Interpreting S_{12} may be extremely difficult, especially when p_1 or p_2 is large.

Let us consider two linear combinations

$$
\begin{aligned}
Y_{11} &= \alpha_1^{\mathrm{T}} X_1 = \alpha_{11} X_{11} + \alpha_{12} X_{12} + \ldots + \alpha_{1p_1} X_{1p_1} \\
Y_{21} &= \beta_1^{\mathrm{T}} X_2 = \beta_{11} X_{21} + \beta_{12} X_{22} + \ldots + \beta_{1p_2} X_{2p_2} .
\end{aligned}
\tag{5.1}
$$

CCA aims at estimating α_1 and β_1 so that Y_{11} and Y_{12} have the *largest* possible correlation. The resulting linear combinations are the first *canonical variables*. By paralleling PCA, the *first* canonical variables will contain the largest amount of total correlation that any linear combination between X_1 and X_2 can summarize. Then, we can proceed with the *second* set of canonical variables Y_{12} and Y_{22}, and so on.

Let us clarify some issues with notation here: the jth linear combination of X_1 is denoted by Y_{1j}, the jth linear combination of X_2 is denoted by Y_{2j}. These are obtained through coefficients α_j and β_j, respectively.

The first pair of canonical variables is obtained by solving the maximization problem

$$
(\hat{\alpha}_1, \hat{\beta}_1) = \mathrm{argmax}_{(\alpha,\beta)} \rho(\alpha, \beta)
\tag{5.2}
$$

with

$$
\rho(\alpha, \beta) = \frac{\alpha^{\mathrm{T}} S_{12} \beta}{\sqrt{\left(\alpha^{\mathrm{T}} S_{11} \alpha\right)\left(\beta^{\mathrm{T}} S_{22} \beta\right)}} .
\tag{5.3}
$$

The measure (5.3) above is nothing but the Pearson correlation coefficient between the linear combinations by fixing α and β. Since the correlation coefficient (5.3) is invariant under scaling of the vectors α and β, a normalization constraint is introduced for identifiability. This corresponds to requiring that Y_{11} and Y_{21} have unit variance, or alternatively that α and β have unit norm.

The value $\hat{\rho}_1 = \rho(\hat{\alpha}_1, \hat{\beta}_1)$ is the first *canonical correlation*, based on coefficients or *canonical weights* $\hat{\alpha}_1$ and $\hat{\beta}_1$. Higher order canonical variables (Y_{1j}, Y_{2j}) can, then, be obtained similarly, with the additional constraint that the jth canonical variables are not correlated with any other canonical variable previously computed. Formally, the jth canonical coefficient $\hat{\rho}_j$ is the optimum of (5.2) under the constraint that Y_{1j} and Y_{2j} are uncorrelated with the previous ones. It is straightforward to show that the jth canonical correlation will be smaller than or equal to the $(j-1)^{th}$, given that additional constraints are used for the same objective function and same data.

The maximum number of uncorrelated canonical variables is $p = \min(p_1, p_2)$. The canonical correlation coefficients will be ordered as

$$
\hat{\rho}_1 \geq \hat{\rho}_2 \geq \ldots \geq \hat{\rho}_j \geq \ldots \geq \hat{\rho}_p.
$$

A quick solution to the problem of finding the $2p$ canonical variables is given by the fact that the canonical correlation coefficients correspond to the square root of the first p eigenvalues of the product matrix

$$
S_{11}^{-1} S_{12} S_{22}^{-1} S_{21}.
\tag{5.4}
$$

It can be noted that the first p eigenvalues of (5.4) coincide with the first p eigenvalues of

$$S_{22}^{-1} S_{21} S_{11}^{-1} S_{12} \, , \tag{5.5}$$

therefore the problem is symmetric. Eigen analysis of the product matrices above is an efficient means of computing the canonical variables: the eigenvectors of either matrix correspond to $\hat{\alpha}_j$ and $\hat{\beta}_j$, $j = 1, 2, \ldots, p$. They are related through the identities:

$$\hat{\alpha}_j \;=\; \frac{S_{11}^{-1} S_{12} \hat{\beta}_j}{\hat{\rho}_j}$$

$$\hat{\beta}_j \;=\; \frac{S_{22}^{-1} S_{21} \hat{\alpha}_j}{\hat{\rho}_j} \, , j = 1, 2, \ldots, p \, .$$

The canonical correlations can also be computed via an eigen analysis of the *symmetric* matrix

$$S_{11}^{-1/2} S_{12} S_{22}^{-1} S_{21} S_{11}^{-1/2}. \tag{5.6}$$

The canonical correlations correspond to the first p eigenvalues of (5.6). In (5.6), $S_{11}^{-1/2}$ can be easily obtained from the eigen decomposition of S_{11}. The product matrix (5.6) is obtained from that in expression (5.4) by simple matrix algebra. Use of symmetric matrices may be preferred since their eigenvectors can be computed with efficient algorithms.

As well as in the PCA framework, when the variables are characterized by different scales, it is recommended to work with the sample correlation matrix rather than the covariance matrix. In the case of CCA, the canonical correlations will be the same but the canonical variables might be substantially different.

CCA can be performed in R via the function `cancor`, but also through the function `cc` in the library CCA. The function `cc` provides supplemental numerical and graphical tools. Another possibility is given by the function `CCorA` in package vegan. Finally, CCA can be easily implemented by using basic commands for matrix algebra, such as `solve`, `eigen` and `svd`.

5.1.1 Interpretation of the results

Interpretation of the canonical variables is based on the canonical weights α and β, in the same fashion as PCA. The eigenvectors of the product matrices (5.4), (5.5) or (5.6) are interpreted according to the rule that relatively large coefficients (in absolute value) correspond to more important measurements. Variables whose coefficients are of opposite signs are contrasted in the same fashion as PCA. Some examples of interpretation in real situations will be given at the end of the chapter.

A different approach uses the *canonical loadings*, which measure the linear correlation between the jth canonical variable and each component variable in one set. We stress that only in the case of CCA the term loadings refers to correlations between variables and linear combinations, rather than to their coefficients. Once again, larger loadings correspond to more important variables in determining the association of interest. Provided that the canonical coefficients from (5.4), (5.5) or (5.6) have been scaled so that the canonical scores have unit variance, the vector of canonical loadings for the jth canonical variable and its component variables can be evaluated by using the sub-matrices S_{11} and S_{22} of the *correlation* matrix S:

$$
\begin{aligned}
\hat{r}_{Y_{1j},X_1} &= \mathrm{Corr}(Y_{1j}, X_1) = \hat{\alpha}_j^{\mathrm{T}} S_{11} \\
\hat{r}_{Y_{2j},X_2} &= \mathrm{Corr}(Y_{2j}, X_2) = \hat{\beta}_j^{\mathrm{T}} S_{22} \ .
\end{aligned}
\tag{5.7}
$$

Finally, direct information about the strength of the associations between the variables from different sets can be obtained through the *cross loadings*. Cross loadings measure the linear correlation between the variables from one set and the jth canonical variable from the other. In case of unit variance scores, cross loadings can be evaluated by

$$
\begin{aligned}
\hat{r}_{Y_{1j},X_1}^c &= \mathrm{Corr}(Y_{1j}, X_2) = \hat{\alpha}_j^{\mathrm{T}} S_{12} \\
\hat{r}_{Y_{2j},X_2}^c &= \mathrm{Corr}(Y_{2j}, X_1) = \hat{\beta}_j^{\mathrm{T}} S_{21} \ .
\end{aligned}
$$

5.1.2   Selection of the number of canonical variables

The object of CCA is to summarize the high dimensional relationship between the two sets of variables through $q < p$ linear combinations and their correlations. As usual with dimensionality reduction, an open issue is to select q as small as possible with reasonable information loss. In order to choose the number of canonical variables to retain, one could work with the percentage of *explained* variance and that of *shared* variance.

A squared loading can be interpreted as the proportion of variance of each variable in one set accounted for by the canonical variable from the same set. Therefore, the average of the squared loadings is a measure of the proportion of total variance in one set explained by the jth canonical variable (*explained variance*), that is

$$
\hat{r}_{X_h|Y_{hj}}^2 = \frac{||\hat{r}_{Y_{hj},X_h}||^2}{p_h} \ , h = 1, 2 \ .
\tag{5.8}
$$

On the other hand, the squared canonical correlation $\hat{\rho}_j^2$ gives the amount of variance of the jth canonical variable from one set accounted for by the corresponding canonical variable from the other set. Namely, this is the amount of variance shared by the two canonical variables (*shared variance*).

A measure of overall association between the two sets is the so called *redundancy index* (Stewart and Love, 1968), which gives the total percentage of variance in one set accounted for by the variables in the other set. The redundancy index is defined as

$$R^2_{X_h|Y_{\bar{h}}} = \sum_{j=1}^{q} \hat{\rho}_j^2 \hat{r}^2_{X_h|Y_{hj}} \ , h = 1, 2, \bar{h} = 2, 1 \ . \tag{5.9}$$

The logic behind the index is to sum the shared variances taking into account the fact that the variables in one set can only explain at most the amount of shared variance in the other set, that is,

$$Y_{2j} \xrightarrow{\text{shared variance}} Y_{1j} \xrightarrow{\text{explained variance}} X_1 \ .$$

It can be demonstrated that the redundancy index (5.9) can be also conveniently obtained from the average of squared cross loadings. The value of q can be as usual set so that the redundancy index is large enough, trading off the number of components q. For more details on the topic we refer the reader to Dillon and Goldstein (1984) and Rencher and Christensen (2012).

5.2 CCA based on robust covariance estimation

The lack of robustness of the sample covariance (correlation) matrix makes classical CCA sensitive to the occurrence of outliers. A first attempt to investigate the robustness properties of CCA is due to Romanazzi (1992), who derived the influence functions of classical estimators.

A straightforward and appealing approach to robust CCA, similarly to what occurs in robust PCA, is based on direct use of robust estimates of the covariance/correlation matrix: S, MM, RMCD, SD, OGK estimators can be used to replace the classical covariance matrix. Each of those estimates leads to a different CCA output, and as usual a comparison of at least two or three choices to check sensitivity to the estimator being used is recommended.

The robustness properties of CCA based on robust estimates of the covariance matrix have been studied in Taskinen *et al.* (2006), who derived the influence functions for CCA functionals and found that they directly depend on the influence function of the covariance matrix estimator.

It is important to underline that the canonical vectors based on robust covariance estimates do not have a natural interpretation anymore, given they do *not* maximize the linear correlation among robust canonical variates. See Branco *et al.* (2005) on this issue. Furthermore, loadings and cross loadings can not be evaluated by using the Pearson correlation coefficient anymore, since it is not robust. An alternative to Pearson correlation is given by pairwise robust measures of correlation. These correspond to robust estimates of two-dimensional correlation matrices, and can therefore be based on truncation, rejection, bounded ψ-functions or any robust covariance estimator. A simpler approach is to rely on Spearman's rank correlation or on the robust index suggested by Gnanadesikan and Kettenring (1972). See Maronna *et al.* (2006), p. 204–208, for a more detailed account and Shevlyakov and Smirnov (2011) for a recent survey on robust correlation.

5.3   Other methods

Instead of using a robust covariance estimate, as well as in the PCA framework, we may project the data at hand on the unit sphere. This usually mitigates the effect of outliers (Dehon *et al.*, 2000). CCA could then be based on the sample covariance matrix evaluated on the transformed data (along the lines of Section 3.4). In a similar fashion, we can use the covariance based on the rank of an observation x_i, that is defined as

$$\text{rank}(x_i) = \frac{1}{n} \sum_{j=1}^{n} \frac{(x_i - x_j)}{||x_i - x_j||} \ .$$

This leads to *spherical* CCA (SCCA).

   An alternative to robust estimation of the covariance (correlation) matrix is given by projection pursuit. As usual, PP might be employed in order to avoid computing and storing large dimensional covariance matrices. Robust CCA may be obtained with PP by replacing (5.3) with a robust correlation index, and sequentially computing $q << p$ canonical variables. A remarkable feature of the PP approach to CCA is that robust canonical variables still preserve their natural interpretation. This approach can be implemented using functions `ccaProj` and `CCAproj` available from the R package `ccaPP`.

5.4   Examples

The following real data examples are used to illustrate both classical and robust CCA.

5.4.1   Linnerud data

The Linnerud data (Tenenhaus, 1998) contain $n = 20$ measurements of three body variables (`weight`, `waist`, `pulse`) and of performance on three exercises (`pull-ups`, `bendings`, `jumps`). The data have been also analyzed in Pison and Van Aelst (2004) and Taskinen *et al.* (2006) and are available from the R package `mixOmics`.

TABLE 5.1
Linnerud data. Canonical correlation and canonical vectors based on classical, RMCD and MM estimates of covariance.

			Classical	RMCD	MM
		$\hat{\rho}_1$	0.796	0.796	0.826
	weight	$\hat{\alpha}_{11}$	-0.440	-0.474	-0.411
X_1	waist	$\hat{\alpha}_{21}$	0.897	0.869	0.908
	pulse	$\hat{\alpha}_{13}$	-0.034	-0.142	-0.082
	pull-ups	$\hat{\beta}_{11}$	-0.264	-0.053	-0.085
X_2	bendings	$\hat{\beta}_{21}$	-0.798	-0.982	-0.933
	jumps	$\hat{\beta}_{31}$	0.542	0.179	0.350

We compare the results of classical CCA with its robust counterparts based on different robust covariance estimates, namely the RMCD with 25% BP and the MM estimator with 50% BP and 95% shape efficiency. Robust analyses have been performed after standardizing the observed variables by their MAD, whereas classical CCA is based on the correlation matrix (that is, variables have been standardized using the standard deviation).

Table 5.1 gives the first canonical correlation and the corresponding canonical coefficients. The robust analyses give results similar to classical CCA, with slight differences in the value of the coefficients but not in their ordering. The first canonical function is mainly determined by `waist` and `bendings`. The importance of `bendings` compared to the other variables in the X_2 set is more evident after the use of robust methods: its canonical weight grows, while the other weights are remarkably smaller.

Table 5.2 shows the classical and RMCD based robust loadings and cross loadings. In the latter case, pairwise robust correlations have been evaluated by using two different techniques. The values in the third and fourth columns have been obtained by computing bivariate RMCD estimates of the covariance matrices of Y_{1h} and X_{hj}, $h = 1, 2, j = 1, 2, 3$. Then, these estimates have been transformed to correlation matrices through the R function `cov2cor`. The values in the fifth and sixth columns have been obtained through the Huber ψ-function with tuning constant set to 1.345. These correlations are obtained as:

$$\frac{\text{ave}[\psi_H(u)\psi_H(v)[}{\sqrt{\text{ave}[\psi_H(u)^2]\text{ave}[\psi_H(v)^2]}} ,$$

where u and v denote the standardized data and the scores, respectively. We observe that the two methods lead to very similar results.

Robust loadings lead us to conclude that `waist` and `bendings` are very important in establishing a link between the two sets of variables. This confirms the results based on the classical analysis. The loadings suggest that large values of `weight` and `waist` are expected in association with light efforts, whereas `pulse` is positively associated with the second set of variables which measure

TABLE 5.2
Linnerud data. Loadings and cross loadings for classical and robust CCA based on RMCD. Pairwise correlations have been evaluated both by bivariate RMCD and Huber function.

		Classical		Robust			
				bivariate RMCD		Huber	
		Y_{11}	Y_{21}	Y_{11}	Y_{21}	Y_{11}	Y_{21}
X_1	weight	0.621	0.494	0.706	0.510	0.594	0.469
	waist	0.925	0.737	0.955	0.689	0.909	0.718
	pulse	-0.333	-0.265	-0.444	-0.243	-0.348	-0.219
X_2	pull-ups	-0.579	-0.728	-0.571	-0.714	-0.583	-0.714
	bendings	-0.651	-0.818	-0.655	-0.990	-0.715	-0.990
	jumps	-0.129	-0.162	-0.145	-0.565	-0.173	-0.620

the different exercises. This inverse relation is more clear when looking at the cross loadings. Variables `weight` and `waist` are positively correlated with Y_{21}, which is characterized by a negative coefficient for the dominating variable `bendings`. The latter is negatively correlated with Y_{11}, which is characterized by a positive coefficient for `waist`.

The differences between the classical and robust approaches do not lead to qualitatively different conclusions, but are somehow substantial quantitatively. These are due to the presence of some outliers. Anomalous values may be identified through the scores of the first pair of robust canonical variables. The scores have been obtained after standardization, that is, we subtracted the RMCD estimate of location and divided by the square root of the RMCD variance.

Figure 5.1 shows the scatterplot of the scores of the first pair of canonical variables based on the classical and RMCD CCA solution. The horizontal axis

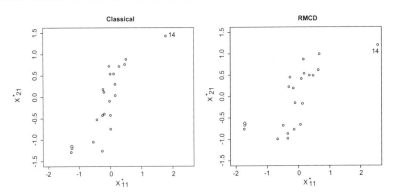

FIGURE 5.1
Linnerud data. First pair of canonical scores based on classical CCA (left) and RMCD (right).

provides an overall index of body measure, the vertical axis an overall index of physical performance. The robust graphical device clearly identifies two points, corresponding to observations 9 and 14, that lie far from the others. These points violate the general correlation pattern. The two data points both fall in the RMCD trimmed set and also receive a low weight during MM estimation. These points are extreme in some of the measurements, which explains the results. Observation 9 has the lowest measurement for `waist` recorded in the data, and the largest one for `pulse`. Therefore, it probably corresponds to someone who is out of training and for whom any moderate physical activity leads to a faster heartbeat. Observation 14 has the largest values for `weight` and `waist`, but the lowest for `pull-ups` and `bendings`. Her/his physical performance is worse than can be expected when comparing with the rest of the data. This happens probably because the subject is overweight and out of training.

Further insight on the presence of outliers is given by a scatterplot of the scores of the first pair of canonical variables and of each variable from both sets. This is reported in Figure 5.2, where observations 9 and 14 are once again marked as outlying. In addition, also observation 10 stands out, especially with respect to Y_{21} and variable `jumps`. This outlier is masked in Figure 5.1. Observation 10 corresponds to someone who is highly fit, with an extreme number of jumps recorded.

We conclude by illustrating spherical CCA for the Linnerud data. SCCA leads to a first canonical correlation equal to 0.907. Canonical coefficients and canonical loadings have the same interpretation as before. The canonical scores are in Figure 5.3. This scatterplot leads to identify observations 10 and 14 as suspicious. On the other hand, observation 9 lies well inside the .975-level tolerance ellipse based on the scaled Beta distribution. The covariance matrix of the projected data has been obtained by using the R function `PcaLocantore`.

5.1 R Illustration (CCA of Linnerud data) *Robust CCA can be carried out with the code that follows. First, after loading the data, we obtain the RMCD fit and build the product matrices (5.4) and (5.5). Here we use the RMCD described in Hardin and Rocke (2004). Any other robust estimate of multivariate location and covariance can be used similarly.*

```
> library(mixOmics)
> data(linnerud)
> x<-linnerud$ex
> y<-linnerud$ph
> xx<-sapply(1:3,function(j) x[,j]/mad(x[,j]))
> yy<-sapply(1:3,function(j) y[,j]/mad(y[,j]))
> est<-CovMcdF(cbind(xx,yy),alpha=0.75)
> Sxx<-est@cov[1:3,1:3]
> Sxy<-est@cov[1:3,4:6]
```

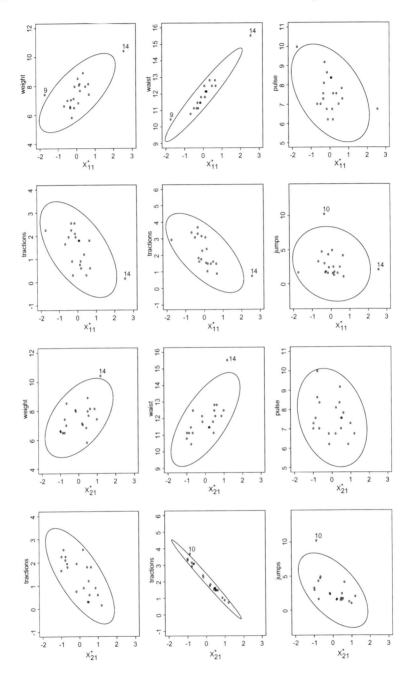

FIGURE 5.2
Linnerud data. Scatterplot of the canonical and original variables. Robust
97.5% tolerance ellipses based on pairwise RMCD.

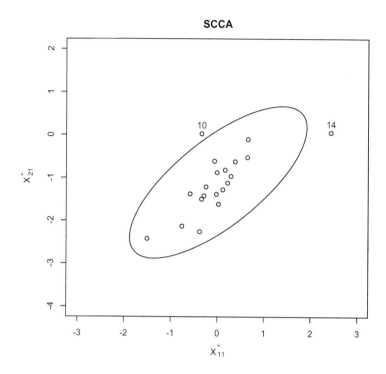

FIGURE 5.3
Linnerud data. First pair of canonical scores based on SCCA with 97.5% tolerance ellipse.

```
> Syy<-est@cov[4:6,4:6]
> Pxx<-solve(Sxx)%*%Sxy%*%solve(Syy)%*%t(Sxy)
> Pyy<-solve(Syy)%*%t(Sxy)%*%solve(Sxx)%*%Sxy
```

We then evaluate the canonical weights and canonical correlations via an eigen analysis.

```
> foralpha<-eigen(Pxx)
> forbeta<-eigen(Pyy)
> rho<-sqrt(forArmcd$values)
> alpha<-foralpha$vector
> beta<-forbeta$vector
```

Finally, the scores are computed as follows:

```
> uxx<-sapply(1:3,function(j) (xx[,j]-est@center[j])/
+ sqrt(est@cov[j,j]))
```

```
> vxx<-uxx%*%alpha
> uyy<-sapply(4:6,function(j) (yy[,j]-est@center[j])/
+ sqrt(est@cov[j,j]))
> vyy<-uyy%*%beta
```

5.4.2   Butterfly data

The butterfly data concerns a study of 16 colonies of the butterfly *Euphydryas edhita* in California and Oregon. There are two types of variables: four environmental variables (`altitude`, annual `precipitation`, minimum and maximum `temperature`) and six genetic variables regarding allelic frequencies for different genes coding phosphoglucose-isomerase (Pgi) enzymes (McKechnie *et al.* (1975), Manly (2005)). Genes with different alleles code proteins with tiny differences. Allelic frequencies can be used to establish how different the butterfly colonies are, and possibly also to understand the sequence of evolution events that has been brought from one kind of colony to another. Additionally, assessment of heterogeneity (that is, evaluation of the entropy or similar measures of variability of allelic frequencies) is of interest *per se* in genomic analysis. Data are stored in R within library `ade4`.

The aim of our example is to relate the Pgi frequencies to environment. This can be useful in order to identify migration and selection effects. Selection corresponds to strong dependence between Pgi frequencies and environmental variables. More precisely, (adverse) environmental conditions may lead individuals with certain alleles to die, therefore leading to a low or even zero frequency for certain Pgi configurations. The same allele might be recorded within colonies living in different environments.

Data have been processed by aggregating the 0.4 and 0.6 allelic frequencies and omitting the 1.30 frequency.

Robust CCA has been based on the 50% BP and 95% shape efficient MM estimator. The distance-distance plot in Figure 5.4 is based on the MM estimate of covariance. It clearly flags colonies `GH` and `GL` as outliers. These anomalous colonies were otherwise masked in the classical analysis. Also the colonies `SB`, `MC` and `DP` are characterized by large robust distances. It can be seen from raw measurements that the former two colonies are characterized by altitude and temperature values noticeably different from the other species. The `DP` colony might be outlying because of its measurement of 1-Pgi frequency, as will be better illustrated below.

Classical and robust correlation structures are displayed in Figure 5.5. The 97.5% level tolerance ellipses are based on the χ^2_2 approximation. Remarkable differences are observed, and they are due to the outliers.

The robust correlations also indicate that weaker relationships among allelic frequencies and environmental variables exist, with respect to those based on the classical analysis. The seemingly strong relationship between genes and environment, indicating dramatic selection effects, might be slightly

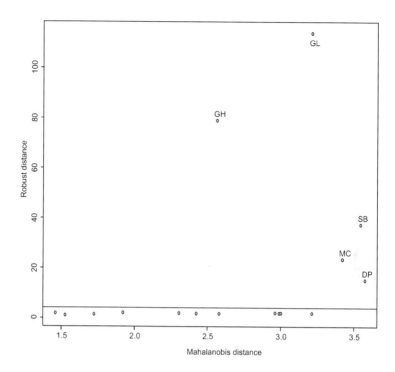

FIGURE 5.4
Butterfly data. Distance-distance plot based on MM.

milder. Classical CCA over estimates the effects of natural selection due to
the presence of the extremal outliers identified previously. Robust CCA gives
a stronger role to migration effects.

In correspondence with the two largest 1-Pgi frequency values we observe
two atypical values for `altitude` and `maximum temperature`, corresponding
to colonies GH and GL. The habitat of these colonies is rather peculiar. These
colonies lead to shifted classical tolerance ellipses, as can be seen in the corresponding panels in the seventh column of Figure 5.5. The strength of the linear
relationship between the allelic frequencies 0.8 and 1.6 and the environmental
characteristics is also biased. Three robust correlations are almost zero, indicating that Pearson's coefficients were completely determined by colonies GH
and GL. This is also testified by the robust fit shown in the first panels of the
last column of Figure 5.5.

Classical and robust CCA are carried out based on the correlation matrices
in Figure 5.5. The robust correlation matrix is based on $\hat{\Sigma}_{MM}$. Both the classical and robust canonical weights, together with canonical correlations, are
reported in Table 5.3. The main difference corresponds to the 1-Pgi frequency.

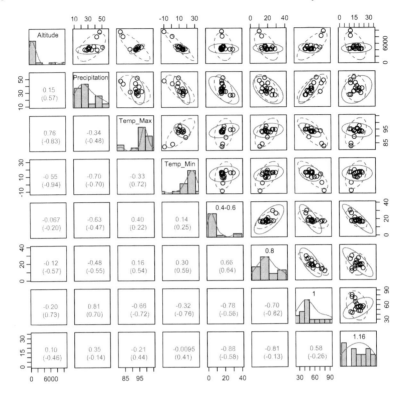

FIGURE 5.5
Butterfly data. Histograms, correlation matrix and pairwise scatterplots with
classical (dashed line) and robust (solid line) 97.5% tolerance ellipses.

We now compute the correlations between the original and canonical variables. These are given in Table 5.4. Robust loadings have been obtained
through both Spearman's rank correlation coefficient and Huber ψ-function.
Even with some slight differences among them, the loadings and cross loadings
based on both approaches suggest that Y_{11} is best interpreted as a measure
of high maximum temperatures and low precipitation. The variable altitude
is much less relevant as suggested by the classical CCA. On the other hand,
the canonical variable Y_{12} is basically a measure of lack of 1-Pgi frequency.
Accordingly, 1-Pgi allele might be seen as an allele that has undergone natural
selection. More precisely, 1-Pgi frequencies are positively associated with low
temperatures and heavy precipitations. Dry and hot environments are probably not amenable to Pgi enzymes coded by genes with 1 allele. This finding
should be better investigated as there are outlying colonies somehow violating
this claim. This might be due to some other environmental or genetic characteristics that are not measured. For instance, a hot and dry but breezy climate

TABLE 5.3
Butterfly data. Canonical coefficients of the first pair of canonical components and first canonical correlation.

	Classical	Robust
$\hat{\rho}_1$	0.862	0.994
Altitude	-0.200	-0.044
Precipitation	-0.471	0.351
Max temp.	0.752	0.789
Min temp.	0.417	0.503
0.4-0.6	0.507	-0.541
0.8	0.390	-0.371
1	-0.082	-0.607
1.16	0.764	-0.449

might reduce the selection effect. More likely, there might be other genes that interact with the Pgi genes and have different alleles as well.

Figure 5.6 reports the scores of the first pair of canonical variables based on the classical and robust analyses. The 97.5% tolerance ellipse in the right panel is based on the 50% BP and 95% shape efficient bivariate MM estimator. The right panel flags, in accordance with the distance-distance plot in Figure 5.4, colonies GH, GL, DP, MC and SB. More in detail, GH and GL have low scores for both canonical variables. They are mainly characterized by low temperatures and high 1-Pgi frequencies. For SB, we have a discordance between the observed temperatures and 1-Pgi frequency. The overall trend is to have a large 1-Pgi with high temperatures, while for colony SB we have a large 1-Pgi corresponding to slightly lower maximum temperature. Colonies DP and MC are also located far from the rest of the data cloud. For DP we observe a large 1-Pgi frequency but low precipitation, leading to a small score on Y_{21}. For MC

TABLE 5.4
Butterfly data. Loadings and cross loadings for classical and robust CCA. The latter are based on Spearman's rank correlation index and Huber ψ-function.

	Classical		Robust			
			Spearman		Huber	
	Y_{11}	Y_{21}	Y_{11}	Y_{21}	Y_{11}	Y_{21}
Altitude	-0.921	-0.794	0.028	-0.131	-0.196	-0.448
Precipitation	-0.771	-0.664	-0.301	-0.610	-0.464	-0.664
Max temp.	0.898	0.774	0.914	0.617	0.927	0.624
Min temp.	0.919	0.792	0.271	0.626	0.210	0.619
0.4-0.6	0.331	0.384	0.383	0.551	0.242	0.424
0.8	0.637	0.740	0.413	0.669	0.540	0.678
1	-0.828	0.961	-0.625	-0.904	-0.586	-0.878
1.16	0.410	0.475	0.261	0.237	0.463	0.474

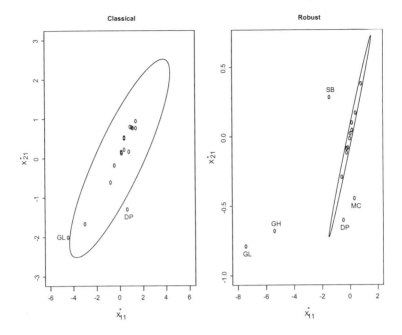

FIGURE 5.6
Butterfly data. Scores of the first pair of canonical variables for classical and robust MM estimates of the correlation matrix. 97.5% tolerance ellipses based on MM.

we observe the largest precipitation measurement, a large 1-Pgi frequency but also a large maximum temperature.

We conclude with CCA based on projection pursuit. In order to perform projection pursuit, we need to select a robust bivariate estimator of correlation. In our example we used the Spearman correlation coefficient and Huber estimator. The two procedures lead to values that are very close for what concerns the first canonical correlation, that is, 0.838 and 0.811 respectively.

5.2 R Illustration (CCA of butterfly data) *CCA based on the Spearman correlation coefficient and on Huber estimator can be obtained with the following code:*

```
> library(ccaPP)
> outSp<-ccaProj(x,y, k = 1, method = "spearman")
> outM<-ccaProj(x,y, k = 1, method = "M")
```

6

Factor Analysis

This chapter deals with robust approaches to Factor Analysis. Factor Analysis (FA) is closely related to PCA, but the reasoning is rather different. The most important difference is that PCA follows from purely geometrical considerations, while Factor Analysis is based on an underlying statistical model. Statistical modeling brings about a set of inferential tools (e.g., hypothesis testing, confidence intervals).

FA *models* the covariance/correlation structure of a p-dimensional set of variables as a function of a smaller number of *latent* variables. These are the *factors*, obtained as linear combinations of the starting variables. Factor Analysis is extensively used in psychological, behavioral, health and social sciences (Thompson, 2004; Cureton and D'Agostino, 2013) but also in econometrics (Croux and Exterkate, 2011) and geostatistics (Filzmoser, 1999).

Classical estimation techniques rely on the sample covariance (correlation) matrix and, hence, the results can be badly influenced by the presence of outliers. Robust FA, much like robust PCA, is obtained through robust estimates of multivariate location and covariance (Pison *et al.*, 2003).

6.1 The FA model

Factor Analysis (FA) is designed to summarize the covariance structure of p variables with $q < p$ *factors*. Factors are interpreted as common latent traits. These traits are *latent* given that we do not observe them directly, and *common* given that they are indirectly measured by the p variables. For this reason, they contain information that is common to the p variables.

A typical example is given by FA of questionnaires. This is often used to measure an attitude (and attitude related aspects), such as, for instance, mathematical or verbal abilities. Questionnaires are based on a large number p of items. The performance for each item is recorded. Mathematical ability can be seen as the single common latent trait of interest, so if the questionnaire is well calibrated (i.e., it measures *only* mathematical ability) we can set $q = 1$. The resulting factor will be a summary measure of mathematical ability. Other typical latent traits are intelligence or other abilities, satisfaction, psychological tendencies and other hypothetical constructs, that, at least in principle, can not be measured directly.

According to the FA model, each variable can be expressed as a linear combination of q factors, plus an additional idiosyncratic error term. The error term is idiosyncratic as the relationship holds only approximately. The classical FA model is

$$X_j = \sum_{s=1}^{q} a_{js}F_s + e_j, \ j = 1, 2, \ldots, p \ . \tag{6.1}$$

The coefficient a_{js} is the loading of the jth variable on the sth factor and F_s is the sth common (unobservable) factor. Let $\mathbf{X} = (X_1, X_2, \ldots, X_p)^{\mathrm{T}}$ be zero centered. The model can be expressed in the matrix notation

$$\mathbf{X} = AF + E, \tag{6.2}$$

where $A = [a_1, a_2, \ldots, a_q]$ is the $p \times q$ matrix of *factor loadings*, $F = (F_1, F_2, \ldots, F_q)^{\mathrm{T}}$ and $E = (e_1, e_2, \ldots, e_p)^{\mathrm{T}}$. We further assume that:

1. the factors are zero centered and their covariance matrix is the identity matrix of dimension q, i.e. I_q;

2. the residuals are also zero centered and their covariance is $\Psi = \mathrm{diag}(\psi_j)$, where the ψ_j are the so-called *uniqueness* or *specific variances*;

3. factors and errors are independent.

The model defined by expression (6.2) and the above assumptions 1-2-3 is also known as the *orthogonal* factor model (Johnson and Wichern, 2002; Manly, 2005). In fact, it is assumed (and therefore imposed at the estimation stage) that factors have zero correlation.

The orthogonal factor model implies that the covariance matrix of $\mathbf{X}$ is

$$\Sigma = AA^{\mathrm{T}} + \Psi \ . \tag{6.3}$$

According to (6.3), the variance of X_j is the sum of two components:

$$\sigma^2_{X_j} = \Sigma_{jj} = h_j^2 + \psi_j \ , j = 1, 2, \ldots, p$$

where the first term $h_j^2 = \sum_{s=1}^{q} a_{js}^2$ is the *communality* and can be interpreted as the amount of variance of X_j accounted for by the q common factors.

Under the orthogonal FA model we have that $\mathrm{Cov}(\mathbf{X}, F) = A$, hence the loading of the jth variable for the sth factor a_{js}, $j = 1, 2, \ldots, p$, $s = 1, 2, \ldots, q$ gives a measure of the covariance between X_j and F_s. Communalities and loadings can be used for interpreting and labeling the factors.

Note that there is not a unique set of loadings, since the matrix A is determined right up to multiplication by an orthogonal matrix. This ambiguity can be solved by various *rotation* techniques, or better by imposing additional constraints, for instance requiring that the matrix $A^{\mathrm{T}}\Psi^{-1}A$ is diagonal (see Johnson and Wichern (2002) for a detailed account on the topic).

It can be demonstrated that the maximum number of factors we can identify is the integer part of

$$q_{max} = p + 0.5 - \sqrt{2p + 0.25} \ .$$

6.1.1 Fitting the FA model

In the classical approach to FA, estimates of A and Ψ are obtained by deconvolution of the sample covariance matrix S according to (6.3):

$$S = \hat{A}\hat{A}^{\mathrm{T}} + \hat{\Psi} \ . \tag{6.4}$$

The sample correlation matrix should be used when the scales of the variables differ substantially, and the procedure is the same.

In order to estimate the classical FA model one can use any of several approaches. Here we describe three of them: principal components, principal factors and maximum likelihood (based on the multivariate normal distribution).

1. **Principal Components**.

    A complete trivial solution to the problem is given by classical PCA. Set

    $$S = \hat{V}\hat{\Lambda}\hat{V}^{\mathrm{T}} = \hat{A}\hat{A}^{\mathrm{T}},$$

    where as usual $\hat{V} = [\hat{v}_1, \hat{v}_2, \dots, \hat{v}_p]$ is the matrix of eigenvectors of S, and $\hat{\Lambda} = \mathrm{diag}(\hat{\lambda}_j)$ is a diagonal matrix whose diagonal contains the corresponding eigenvalues.

    The jth vector of estimated factor loadings $\hat{a}_j$ corresponds to $\sqrt{\hat{\lambda}_j}\hat{v}_j$, $j = 1, 2, \dots, p$.

    This solution is clearly not interesting because we have as many factors as variables, with $\hat{\Psi}$ being null.

    We want to allow for residual error, consequentially performing dimension reduction. The basic idea is that the matrix $\hat{A}$ can be approximated by assuming that the last $p - q$ eigenvalues are equal to zero *at population level*. Therefore, we fix $\hat{A} = [\hat{a}_1, \dots, \hat{a}_q]$, with $\hat{a}_j$ as above. We then compute $\hat{\Psi} = S - \hat{A}\hat{A}^{\mathrm{T}}$. Consequently,

    $$\hat{h}_j^2 = \sum_{s=1}^{q} \hat{a}_{js}^2 = \hat{v}_j^{\mathrm{T}}\hat{\lambda}_j\hat{v}_j = \hat{\lambda}_j \ .$$

    This method recovers the diagonal of S exactly, but not its offdiagonal elements. Goodness of fit will therefore depend on the size of the off-diagonal elements of the residual matrix $\tilde{S} = S - (\hat{A}\hat{A}^{\mathrm{T}} + \hat{\Psi})$. A good fit is obtained when the sum of the squares of the last $p - q$ eigenvalues is small.

2. **Principal Factors.**

Let $\hat{\Psi}^{(0)}$ be a preliminary estimate of Ψ and evaluate the *reduced* sample covariance (or correlation) matrix

$$S^r = S - \hat{\Psi}^{(0)} \ ,$$

obtained by replacing the variances s_{jj} on the main diagonal by the initially fitted communalities $(\hat{h}_j^2)^{(0)} = s_{jj} - \hat{\psi}_j^{(0)}$, $j = 1, 2 \ldots, p$.

The method of principal factors proceeds by estimating $S^{(0)}$ using the first q eigenvalues $\hat{\lambda}_j^{(1)}$ and eigenvectors $\hat{\nu}_j^{(1)}$, of S^r, $j = 1, 2, \ldots, q$, that is

$$S^{(0)} = \hat{A}^{(1)}(A^{(1)})^{\mathrm{T}} + \hat{\Psi}^{(1)}$$

with $A^{(1)} = [\hat{a}_1^{(1)}, \hat{a}_2^{(1)}, \ldots, \hat{a}_q^{(1)}]$ and $\hat{a}_s^{(1)} = \sqrt{\hat{\lambda}_j^{(1)}}\hat{\nu}_j^{(1)}$, $s = 1, 2, \ldots, q$. Then, the communalities are updated as $(\hat{h}_j^2)^{(1)} = \hat{\lambda}_j^{(1)}$ and a new reduced matrix is calculated. The procedure is then iterated until convergence. A good initial estimate is crucial for success of this estimation method. One popular choice is to set $\hat{\psi}_j^{(0)} = 1/s^{jj}$, where s^{jj} is the jth diagonal element of S^{-1}.

3. **Maximum Likelihood.**

The method of maximum likelihood relies on the assumption that data arise from a multivariate normal distribution with covariance matrix (6.3). The estimates $(\hat{A}, \hat{\Psi})$ are obtained by maximizing the log-likelihood function

$$\ell(A, \Psi) \propto \left[-\log |AA^{\mathrm{T}} + \Psi| + tr\left((AA^{\mathrm{T}} + \Psi)^{-1} S \right) \right] \ ,$$

under the constraint that $A^{\mathrm{T}}\Psi^{-1}A$ is diagonal.

It is also worth mentioning the MINRES (Minimum Residuals) method (Harman and Jones, 1966). Factors are estimated by minimizing

$$\sum_{j_1=2}^{p} \sum_{j_2=1}^{j_1-1} \left(s_{j_1 j_2} - \sum_{s=1}^{q} a_{j_1 s} a_{j_2 s} \right)^2 \ .$$

The number of factors q can be chosen on the basis of the proportion of *explained variance* as measured by

$$\frac{\sum_{j=1}^{q} \hat{\lambda}_j}{\sum_{j=1}^{p} s_{jj}} \ ,$$

where the denominator becomes equal to p when the correlation matrix is being used.

Some care is needed with FA model (6.2), given that for some values of q the problem may become ill-conditioned, with negative estimates of the eigenvalues or other inadmissible estimates. Those values for q should obviously be discarded.

There are several other methods to select q: here we mention the standard approaches based on the screeplot and on Kaiser's rule, and the so called *parallel analysis* (Horn, 1965; Hoyle and Duvall, 2004) based on Monte Carlo simulation.

Factor scores play a central role in FA. Scores project data onto a low dimensional space, similarly to scores in PCA. Factor scores can be estimated in FA using for instance one of the two following techniques:

1. **Weighted Least Squares**.

   Factor scores may arise from weighted linear regression, where the weights are given by $\hat{\psi}_j^{-1}$:

   $$\hat{F} = (\hat{A}\hat{\Psi}^{-1}\hat{A}^{\mathrm{T}})\hat{A}\hat{\Psi}^{-1}(x_i - \bar{x}) \tag{6.5}$$

2. **Regression**.

   If we assume (X, F) are jointly normally distributed, then

   $$\hat{F} = \hat{A}^{\mathrm{T}} S^{-1}(x_i - \bar{x}) . \tag{6.6}$$

As with PCA, one can rotate the loadings in order to aid interpretation, if needed. We give here some additional insights into rotation. Rotation is always of the form $\tilde{A} = \hat{A}G$, where G is a suitably chosen rotation matrix. We can distinguish among two type of rotations: orthogonal and oblique (where the orthogonality constraint is relaxed). Among orthogonal rotations, the most popular is the *varimax* (Kaiser, 1958), whereas the *oblimin* is commonly used as an oblique rotation technique. Oblique techniques are a last resort when no orthogonal technique leads to interpretable loadings, but their results should be interpreted with care. We point the reader to Jolliffe (2005) for additional details on rotation methods. Furthermore, sparseness of the factor loadings is an attractive solution in order to facilitate and improve interpretation of the factors (Croux and Exterkate, 2011).

Factor analysis can be performed in R by using the `factanal` function, which is based on maximum likelihood estimation. An alternative is given by function `fa` in library `psych`. The latter includes additional options.

6.2   Robust factor analysis

The most direct and appealing approach to robust FA is to perform the usual procedures but to approximate a robust estimate of the covariance/correlation matrix. The robust estimate shall be used in (6.4), for instance. Such an estimate can be chosen among those illustrated in Chapter 2.

Pison *et al.* (2003) recommended a robust FA based on the RMCD and derived the robustness properties and the influence function of the method. Robust estimation of communalities and specific variances can also be pursued simply by replacing the sample covariance/correlation matrix by its robust counterpart when fitting the FA model.

A different approach has been discussed in Croux *et al.* (2003), who proposed an alternating regression algorithm based on a robustification of the least squares approach to FA.

6.3   Examples

In the following examples, classical FA was performed through the R function `fa` within library `psych`. Robust methods can be obtained by feeding robust estimates to `fa`, or directly using the function `FaCov` in library `robustfa` (see Appendix A.5).

One drawback of the latter function is that rotation techniques, and related communalities adjustments, are not automatically supplied. Rotation should be performed using the `varimax` function in library `stats`, or using one of the several rotation functions in package `GPArotation`.

6.3.1   Automobile data

Let us consider the Automobile data, where we have twenty cars equipped with a diesel engine that are outliers with respect to the rest of the data. In particular, cars running on diesel have a larger compression ratio. These cars stand out in the distance-distance plot in Figure 6.1, which is based on the 25% breakdown point RMCD. Cars with a diesel engine are identified by the symbol +. These outliers seem to be *orthogonal*. They are masked in the classical analysis and only visible when using the robust fit. Here, we used the detection rules outlined in (2.28) and (2.29). The dashed line is based on the scaled Beta distribution and can be used to detect outliers in the non-trimmed set of data (circles), the solid line is based on the scaled F distribution and provides a threshold to identify outliers in the trimmed set (triangles and +).

We fix $q = 3$ factors based on the RMCD correlation matrix $\hat{R}_{RMCD}$,

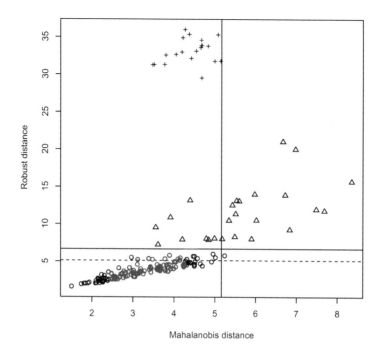

FIGURE 6.1
Automobile data. Distance-distance plot based on 25% BP RMCD.

computed as

$$\hat{R}_{RMCD} = \hat{D}^{-1/2}\hat{\Sigma}_{RMCD}\hat{D}^{-1/2}$$

with $\hat{D} = \text{diag}(\hat{\sigma}_{jj})$ and $\hat{\sigma}_{jj} = \left[\hat{\Sigma}_{RMCD}\right]_{jj}$. We used the principal factors method. Furthermore, the final loadings have been rotated through varimax.

We now compare the results of classical and robust FA. Table 6.1 gives the communalities for each standardized variable and the loadings for the first three factors. We do not observe large differences in the loadings for the first two factors, while, as far as the third factor is concerned, `compression ratio` and `peak-rpm` have a large loading when computing the classical estimate. On the other hand `bore` and `stroke` dominate the robust one. The three classical factors take into account 60% of the variability of `compression-ratio`, but this communality drops to 22% in the robust fit. The high percentage of explained variance is due to the outlying cars. If interpretation is based on the classical estimates, one might wrongly conclude that *compression-ratio* is strongly related to the latent variables. Robust communalities for `bore` and `stroke` are noticeably larger than the classical ones.

TABLE 6.1
Automobile data. Loadings of the first three factors and communalities from classical and RMCD-FA. The last line gives the cumulative percentage of explained (robust) variance.

	Classic				RMCD			
	F_1	F_2	F_3	h^2	F_1	F_2	F_3	h^2
symboling	-0.03	**-0.63**	-0.09	0.40	0.09	**-0.66**	-0.10	0.46
wheel-base	**0.57**	**0.72**	0.25	0.92	**0.54**	**0.81**	0.05	0.95
length	**0.75**	**0.56**	0.17	0.90	**0.73**	**0.54**	0.17	0.86
width	**0.77**	0.36	0.29	0.80	**0.78**	0.37	0.04	0.74
height	0.06	**0.78**	0.06	0.61	0.04	**0.69**	0.18	0.51
curb-weight	**0.89**	0.30	0.27	0.96	**0.94**	0.25	0.11	0.96
engine-size	**0.90**	-0.04	0.30	0.91	**0.88**	0.05	0.02	0.79
bore	**0.63**	0.21	0.06	0.44	**0.62**	0.20	**0.59**	0.77
stroke	0.11	-0.04	0.28	0.09	0.24	-0.06	**-0.65**	0.48
compression-ratio	-0.14	0.25	**0.72**	0.60	**-0.43**	0.17	0.09	0.22
horsepower	**0.93**	-0.19	-0.05	0.90	**0.94**	-0.04	-0.06	0.88
peak-rpm	-0.05	-0.32	**-0.47**	0.33	-0.10	-0.10	-0.33	0.13
city-mpg	**-0.92**	-0.08	0.32	0.95	**-0.94**	-0.04	-0.05	0.90
highway-mpg	**-0.92**	-0.15	0.25	0.93	**-0.92**	-0.06	-0.08	0.85
price	**0.86**	0.05	0.22	0.79	**0.84**	0.20	0.09	0.75
Cumulative var. %	0.45	0.61	0.70		0.47	0.62	0.68	

The first robust factor can be interpreted as a general index of quality of a car. It is basically a weighted average of the measurements. The second factor is a measure of security, and leads us to conclude that big cars have lower actuarial risk. The third factor apparently measures the cubic capacity of an engine cylinder through its bore and stroke.

Robust factor scores were obtained through the regression technique after replacing the sample correlation matrix with $\hat{R}_{RMCD}$ in equation (6.6). Classical and robust biplots are in Figure 6.2. In the biplots, in addition to the scores, we also plot variables with arrows. The orientation of the arrows corresponds to the loadings and their length is proportional to the communality. In summary, in the biplot statistical units and variables are plotted in the same diagram.

The biplots are clearly different both for what concerns the position of the variables and the units.

6.1 R Illustration (Robust FA of automobile data) *Robust FA was performed with the code that follows.*

```
> X<-Automobile[,-2]
> auto.mcd<-CovMcdF(X, alpha=0.75)
> R<-cov2cor(auto.mcd@cov)
> fa.rob<-fa(r=R, nfactors=3,rotate='varimax',
+ covar=FALSE,fm='pa')
```

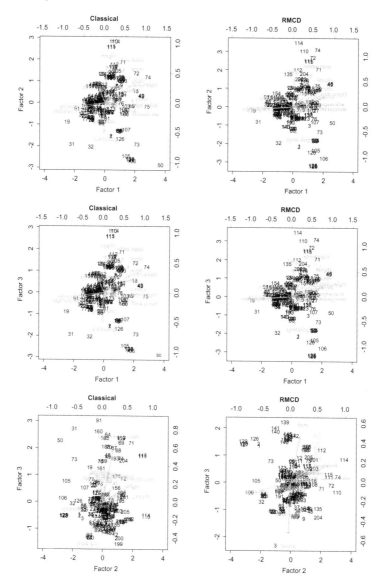

FIGURE 6.2
Automobile data. Classical and RMCD based biplot.

The variables are standardized according to the robust fit and collected in the data matrix Xc . Robust regression scores were computed as follows:

```
> Xc<-sapply(1:ncol(X), function(j)
+ (X[,j]-auto.mcd$center[j])/sqrt(auto.mcd$cov[j,j]))
> hat.f<-t(t(fa.rob$loadings)%*%solve(R)%*%t(Xc))
```

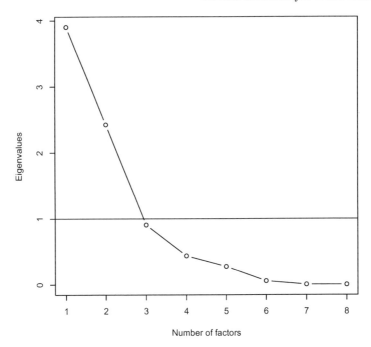

FIGURE 6.3
Butterfly data. Robust screeplot based on MM.

The `biplot` *based on the RMCD was obtained with the following code:*

```
> biplot(x=t(hat.f.rob)[,1:2], y=rob.af$loadings[,1:2],
+ xlab='Factor 1', ylab='Factor 2', main='RMCD',
+ xlim=c(-4,4),col=c('black', 'gray'))
```

This biplot concerns the first two factors, but it can be modified easily to show other factors (e.g., the first and the third).

6.3.2   Butterfly data

Let us now consider the Butterfly data, also described in Section 5.4.2. We proceed with FA for the whole set of variables (both environmental and genetic factors). We have already seen, as shown in Figure 5.4, that some colonies are outlying and this has consequences on covariance/correlation estiamates when robust methods are not used.

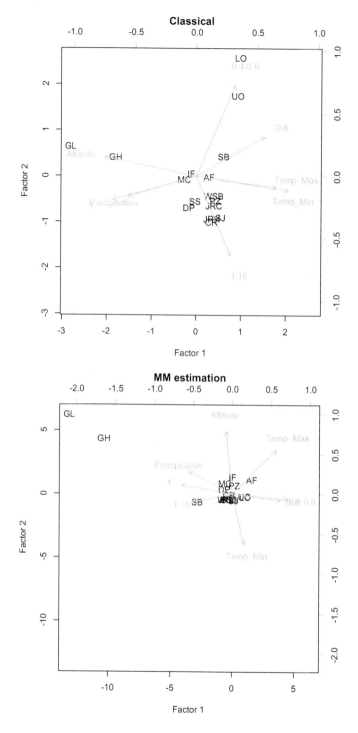

FIGURE 6.4
Butterfly data. Biplots for classical and robust FA based on MM.

The maximum number of factors that can be fit is equal to $q_{max} = 5$. Figure 6.3 shows the screeplot based on MM estimation. Here we decided to keep the first $q = 2$ factors (also according to Kaiser's rule).

The resulting biplots are given in Figure 6.4.

Factors were estimated by maximum likelihood, based on the robust correlation matrix, and then, the varimax rotation was applied to improve the loadings with respect to possibility of interpretation.

Classical and robust biplots show very different interactions between points and variables, due to the different correlations estimated in Figure 5.5. The two outlying colonies GH and GL and the variable altitude are positioned differently when comparing classical and robust plots. Their position in the classical biplot is due to the influence of outliers in the altitude measurements. These correspond to colonies GH and GL.

From the robust biplots it can be seen that the variables precipitation and maximum temperature have large loadings for the first factor, together with the genetic frequencies. Consequently, natural selection has occurred as a consequence of these two environmental conditions. High maximum temperatures and low precipitations appear to be related to lack of 1 and 1.16 Pgi Frequencies, as noted before when using CCA.

Part II

Sample Reduction

Introduction to Sample Reduction

Reducing the number of observations is an important task for describing and understanding the sample at hand. Reducing the number of observations may be needed in order to obtain profiles, which are then used to build policies. The central idea is that heterogeneity in the sample is mostly arising from the presence of very few (say, k) homogeneous groups, as compared to the sample size n. In most cases, this operation is completely unsupervised, that is, groups are not known in advance for any of the subjects. This is the case of *cluster analysis*. In cluster analysis, group labels obtained after cluster analysis are *per se* meaningless. Careful interpretation of cluster profiles helps the user characterize the (optimal) groups. Cluster profiles, to be more carefully defined below, can be seen as the reduced sample and possibly used as a new data set if needed.

When groups labels are known in advance, sample reduction is *supervised* and corresponds to *classification*. In this case, one may simply want to summarize the cluster profiles (possibly using robust measures of location, scatter and any feature of interest). We do not give much detail on this topic further, as robust measures of location and scatter were already discussed in Chapter 2. A different aim of classification could be that of building rules for predicting the class of a future observation. A method for robustly performing this task is briefly reviewed in Chapter 11.

There is a third group of sample reduction techniques that is receiving attention in the machine learning community recently. These techniques are designed for the case in which only a fraction of group labels are known. They are *semi-supervised classification* approaches. These are beyond the scope of this book and we only mention them in these introductory notes. Earlier works about semi-supervised classification include O'Neill (1978) and Ganesalingam and McLachlan (1978), while more recent contributions can be found in Chawla and Karakoulas (2005), Belkin *et al.* (2006). See also Chapelle *et al.* (2006), Liang *et al.* (2007), and references therein. There are still very few robust semi-supervised methods, with few exceptions (e.g., Miller and Browning (2003)).

The next few chapters regard methods for cluster analysis. We begin by introducing the main ideas and briefly reviewing classical non-robust methods for cluster analysis in Chapter 7, we then proceed with robust methods producing spherical clusters in Chapter 8 and robust model based methods (yielding elliptical clusters) in Chapter 9. In Chapter 10 we deal with double clustering, that is, simultaneous grouping of rows and columns of a data

147

matrix. Finally, as already mentioned we outline a robust method for classification in Chapter 11.

7

k-means and Model-Based Clustering

This chapter serves as a very brief overview of classical non-robust methods for sample reduction. We will briefly review the classical k-means procedure, which can be formulated in a non-inferential framework and yields spherical clusters. We will then note that the classical k-means procedure can be cast in an inferential framework and generalized to yield elliptical clusters. The chapter will end with a discussion on choosing the number of clusters.

This chapter is not intended as a thorough review of clustering and model based clustering, which are the objective of research in many fields and have been investigated from very different perspectives. It shall only be intended as the basis material for the robust clustering procedures which will be introduced in the following chapters. For books thoroughly treating classical cluster analysis refer for instance to Mirkin (2005) and Everitt *et al.* (2011), while key references on model based clustering include Fraley and Raftery (2002), McLachlan and Peel (2000) and references therein. We also point the reader to Gan *et al.* (2007).

7.1   A brief overview of applications of cluster analysis

There are many applications of cluster analysis in very different areas. Accordingly, also the typical characteristics and order of magnitude of the data matrix are variable and involve different estimation strategies. We now give few examples as an introduction and overview of the questions that can be answered with cluster analysis.

A popular application of cluster analysis is in genomics, where genes with similar patterns of expression are grouped. There is a huge number of papers describing methods specifically tailored for these applications, see Eisen *et al.* (1998), Jiang *et al.* (2004) and D'Haeseleer (2005). In genomic applications it often happens that subjects correspond to genes, so that the sample size can range from hundreds to tens of thousands. Variables correspond often to repeated measurements (over different subjects, tissues, etc.) of expression, so that they are of the same nature and basically groups may often be summarized by one single (rather than p) measure of location and one of scatter. Furthermore, in this application, groups are often characterized by their level of expression as active or inactive. When measurements are related to a comparison between two biological conditions (e.g., safe and diseased), gene profiles are often characterized as over, under or equally expressed. In this case, measurements represent a log fold change. See also Alfó *et al.* (2007, 2011) and Farcomeni and Arima (2012).

Another popular application is in public health, where promotion of good health policies may start from the identification of subgroups of the population in terms of a multidimensional evaluation of their health status. This is useful in order to identify the specific needs of particular groups of subjects, and tailor an intervention accordingly (e.g., Hodges and Wotring (2000)). Note that interventions may also consist of health promotion campaigns and in education for prevention. For a review see for instance Clatworthy *et al.* (2005).

A third area of application is that of marketing, where cluster analysis is used to identify target markets and for market segmentation. In the former case, cluster analysis is used to profile non-customers, first time buyers, customers and loyal customers, respectively. In the latter case, a target market is divided into subsets of customers with common needs. Market segmentation is also used to identify communication media and strategies most apt to homogeneous subsets of customers. Customers are usually profiled with respect to their behavioral, demographic, psychological and geographical characteristics. Popular reviews and early applications can be found in Frank and Green (1968), Christopher (1969), Saunders (1980).

We conclude by mentioning two emerging areas: text mining and image segmentation. In text analysis, unsupervised learning is used for mining contents in the corpus of documents, and to identify clusters of topics or simply the sentiment (e.g., positive or negative) of a text. In image segmentation, the task is that of decomposition of a gray level or color image into homogeneous tiles (Comaniciu and Meer, 2002). Clusters are often estimated in order to identify borders of objects within the image.

7.2   Basic concepts

Unsupervised classification corresponds to creating groups in such a way that subjects in one cluster are very similar and clearly distinct from subjects in other clusters. The only information used are the p measurements, and no previous information on grouping is available.

A central role is that of an *index of similarity or dissimilarity* between data points. The cluster is made of a group of observations sharing closely related properties *as implied by the dissimilarity index*, that is, they have small mutual distances. Relatedly, a density based definition is given in Carmichael *et al.* (1968): clusters are regions of the measurement space which are densely populated and surrounded by relatively empty regions. A distance is any non-negative function $d(\cdot, \cdot)$ such that $d(x_i, x_j) = 0$ if and only if $x_i = x_j$; which is symmetric $(d(x_i, x_j) = d(x_j, x_i))$; and which obeys to the triangle inequality, that is,

$$d(x_i, x_j) \le d(x_u, x_i) + d(x_u, x_j), \qquad (7.1)$$

for any i, j, u.

The *objective function* summarizes how good a clustering solution is, and its optimizer is the final clustering solution. Its value is almost always nonmeaningful *per se*, but can be compared with other solutions. For instance, in model based clustering the log-likelihood of a constrained and unconstrained solution can be compared in a likelihood-ratio test. A related concept is that of *centroid*. Usually, a compact cluster can be represented by a point or center. This is usually obtained through a location estimator applied to each dimension of the subset of observations within a cluster. The objective function can also often be seen as a measure of the loss of information when replacing the n data points with the k centroids.

7.3   *k*-means

One of the most common clustering algorithms is the k-means procedure. For this procedure, the number of clusters k is fixed in advance and the algorithm proceeds by iteratively finding better and better solutions with respect to an opportune objective function. One of the first instances of the conventional k-means algorithm can be found in Macqueen (1967), later described in more detail by Hartigan (1975) and Hartigan and Wong (1979). There are many k-means-type algorithms, some of which are targeted to improve efficiency for the conventional problem that we describe below, and some of which slightly modify the task.

In the first group, we mention the proposal of Huang (1998), who obtains the optimum by alternating two simpler optimization problems. The method in Huang (1998) is particularly useful when clustering large data sets. Selim and Ismail (1984) perform a comprehensive study of the k-means problem. Tarsitano (2003) presents a survey, with detailed computational comparison, of as many as seventeen variants of the k-means algorithm.

In the second group, we mention k-harmonic means (Zhang *et al.*, 2001), which is designed to be essentially insensitive to the initialization of centroids, and the mean shift algorithm (Cheng (1995), Comaniciu and Meer (2002)). Indeed, also the robust k-means procedures that we will review in the next chapters can be seen as slight modifications of the conventional k-means problem.

We will base our description of the k-means algorithm on the Euclidean distance

$$d(x_i, x_j)^2 = \sum_{h=1}^{p} (x_{ih} - x_{jh})^2. \tag{7.2}$$

Let us also denote the cluster centroid $\bar{x}_c$, $c = 1, \ldots, k$, collected in the k by p matrix $\bar{X}$; and $d_{ic}^2 = d(x_i, \bar{x}_c)$. The optimization problem connected with conventional k-means can be expressed as

$$\min_{\bar{X}} \sum_{i=1}^{n} \min_{c=1,\ldots,k} \sum_{j=1}^{p} (x_{ij} - \bar{x}_{cj})^2, \tag{7.3}$$

where the objective function is therefore obviously

$$\sum_{i=1}^{n} \min_{c=1,\ldots,k} \sum_{j} (x_{ij} - \bar{x}_{cj})^2.$$

The rationale behind (7.3) is readily understood in a sample reduction framework: if we replace the ith observation with its closest centroid, we have an approximation error summarized by $\min_c d_{ic}^2$, which is exactly the kernel of (7.3). The approximation error is then summed over all observations to yield the loss of information when summarizing the n observations with k centroids.

In order to solve (7.3), the algorithm proceeds from an arbitrary initial partition, or equivalently from k arbitrary centroid values $\bar{x}_c$, $c = 1, \ldots, k$. In the first case, $\bar{x}_{cj}$ is simply obtained as the arithmetic average of the jth dimension of the observations initially assigned to cluster c.

At each iteration, the membership of each observation is updated by computing d_{ic}, $c = 1, \ldots, k$ and assigning the ith observation to the closest cluster. For ease of presentation we introduce the $n \times k$ binary matrix U, where $u_{ic} = 1$ if the ith observation is in cluster c and 0 otherwise. Given that all observations are assigned to one and only one cluster, $\sum_c u_{ic} = 1$ and U is a binary matrix. This is particularly convenient as centroid estimates can be directly obtained from the projection matrix derived from U. The conventional iteration of the k-means algorithm is summarized in Algorithm 7.1.

The computational complexity of the algorithm is $O(nkp)$ per iteration. Hence, its complexity is linearly proportional to the size of the data. Since the sequence of objective functions generated is strictly decreasing and the objective function is bounded from below, the algorithm will converge to a local minimum after a finite number of iterations.

A drawback is that the algorithm often yields only a *local* optimum. This is a well known problem, see for instance Anderberg (1973) and Selim and Ismail (1984). The possibility of being trapped in local optima leads to two shortcomings. First of all, *instability*. Two different initial solutions may give different final solutions. Secondly, *non-uniqueness*. The solution of (7.3) is under certain conditions unique. Therefore, when two runs of the algorithm

---

Algorithm 7.1 Conventional k-means algorithm

Update the row memberships
for $i = 1, \ldots, n$ **do**

Let $d_{ic}^2 = \sum_{j=1}^{p} (x_{ij} - \bar{x}_{cj})^2$, $c = 1, \ldots, k$.

Let $r_i = \text{argmin}_c d_{ic}^2$.

Set $u_{ir_i} = 1$. All other elements of the ith row of U are set to 0.

end for

Estimate the centroid matrix
$\bar{X} = (U^T U)^{-1} U^T X$.

---

do not yield the same results, one has been trapped in a sub-optimal solution. Given these considerations, a crucial point is *initialization* of the algorithm. There are different possibilities. One is to randomly assign the cases into k clusters. Another is to choose k cases at random and use them as initial centroids. Better initialization strategies can be found in Khan and Ahmad (2004). Regardless of the initialization strategy, an important recommendation is to repeatedly run the algorithm from many initial solutions, and use the result corresponding to the minimal observed objective function value.

When reached, the global optimum has important properties. It is important to stress that all these properties hold only under the assumption of absence of contamination. Almost sure convergence of the estimated centroids to the population centroids μ_c is established in Pollard (1981), distributional convergence in Pollard (1982). A formal study of the robustness properties of the k-means algorithm follows: the influence function is derived numerically in García-Escudero and Gordaliza (1999). It is demonstrated that the gross error sensitivity is finite, and that the influence function has some discontinuities. Despite the positive result on gross error sensitivity, it is also argued in García-Escudero and Gordaliza (1999) that the universal individual breakdown point is infinitesimal. This result is further strengthened in Farcomeni (2009b), who shows that also the universal cell breakdown point is infinitesimal. Ruwet *et al.* (2013) and Farcomeni (2014b) show the same for an opportunely restricted individual and cell breakdown point, respectively. Hence, according to the results of Ruwet *et al.* (2013) and Farcomeni (2014b), even when the algorithm is run on good data sets with a natural grouping structure, a single outlier and even a single corrupted entry may completely break down the centroid estimates.

We proceed with an illustration of the k-means algorithm, interpretation of its results, and a hint towards its lack of robustness in the following examples.

7.1 Example (Clustering G8 data) *Consider the G8 data relative to macroeconomic performance of eight countries. We standardize the data using* R *function* scale *given that the indicators have a different order of magnitude. Note that standardization itself, not only k-means, may be affected by*

TABLE 7.1
G8 macroeconomic data. Cluster labels obtained with 3-means. Results are based on standardized measurements.

Country	Labels
FRA	3
GER	3
GBR	2
ITA	1
SPA	2
USA	3
JAP	3
CAN	3

the presence of outliers; hence it may be sensible to perform also this step using robust estimates of location and scatter within each of the $p = 7$ measurements. We do not proceed in this way here. After standardization, we use the function `kmeans` to obtain $k = 3$ clusters.

For this small data set we have some instability of the conventional k-means algorithm. This is easily acknowledged as repeated runs of the default `kmeans` function often lead to different results. In order to increase the likelihood of obtaining the global optimum, we use the options `nstart=10` and `iter.max=100`. The results are fairly more stable after specification of these options.

The final cluster labels are given in Table 7.1.

There are two hints when using `kmeans` that there may be outliers affecting the results. One is the relative instability of the conventional k-means algorithm. The second is the presence of a singleton in Table 7.1. Even after standardization Italy is so different from the other seven countries that it forms a group on its own. Even conventional k-means is able to recognize Italy as an outlier. As far as the other countries are concerned, also the other groups obtained may be affected by the presence of outliers, as a second group is made only of Great Britain and Spain. The coefficients of variation in group 2 are small, in particular with respect to GDP (0.82) and interest rates (0.92), indicating that Great Britain and Spain may not belong to the same cluster even if they are assigned to one.

7.2 Example (Clustering metallic oxide data) *Our second example regards the metallic oxide data set. Recall that in this application we have $n = 31$, $p = 8$ and $k = 2$ types of metallic oxide raw material. This is a confirmatory cluster analysis as the true clusters are known in advance (and of course labels are not used for clustering). We compare the true and predicted group labels using the modified Rand index (Hubert and Arabie, 1985), which is a measure of agreement between the true and reconstructed clustering.*

TABLE 7.2

Water treatment plant data. Average of standardized measurements by location and cluster as identified by 3-means.

	Cluster 1	Cluster 2	Cluster 3
Input	-0.07	0.08	-0.08
First Settler	0.02	0.11	-0.23
Second Settler	0.07	0.10	-0.28
Output	0.02	0.01	-0.05

The classical *k*-means procedure with $k = 2$ leads to a very bad solution, in which lots 6 and 7 of Type 2 belong to one group and all the other rows belong to the other group. The resulting Rand index is only 5.8%.

It could be argued that the outliers can be sorted out simply by increasing the number of clusters, but setting $k = 3$ leads to a badly behaved solution as well. It is true that 3-means creates a group made of the two outlying lots (6 and 7 of Type 2), but the remaining lots are not clustered well in terms of type of material. As a matter of fact, the Rand index obtained after exclusion of the two outliers is only about 17%.

7.3 Example (Clustering water treatment data) *Our third example is in regards to a genuine sample reduction problem with the water treatment plant data. We fix $k = 3$ and use k-means. We use direct measurements at input, primary and secondary settlers, and output. We therefore have $p = 29$.*

The sample size of $n = 380$ is reduced to $k = 3$. The three clusters are made of 111, 176 and 93 observations, respectively. The new data set can be further summarized by averaging the $p = 29$ measurements in three groups, as identified by location of measurement. The resulting reduced data set can be found in Table 7.2.

The three clusters can be easily interpreted: the first and third groups identify days of slightly low input loads. For the first group, this is well processed and produces a near to average output. For the third group, the input load is not well processed with bad performance both for the first and second settler. Finally, the second group is made of days with slightly high input, which brings high loads to the first and second settler but no substantial differences in output in comparison to the first group.

At first sight, there is no sign of bias for our results with k-means. We must mention we could obtain alarming figures if we investigated the within cluster variance, or produced cluster specific outlier maps. Nevertheless, without use of those tools or comparison with robust sample reduction methods, we draw our conclusions as above. We are not aware that these are seriously biased by contamination. As we will see in the next chapters, there is strong evidence for the presence of outliers in the data at hand. Robust estimates will lead us to completely different conclusions.

7.4 Model-based clustering

Clustering algorithms can be developed based on probability models, the most important of which is the classical finite mixture model (e.g., McLachlan and Peel (2000)). In our opinion there are mainly two clear advantages and two drawbacks when using model based clustering. The most important advantage is that by using a probability machinery, the user is allowed to perform *inference* on parameters of interest and related clustering. Formal hypothesis testing can be carried out, and the uncertainty associated with sampling can be taken into account and properly summarized with confidence intervals. With some formulations, instead of simple cluster labels, a probability statement can be obtained for the label of each subject. Furthermore, one can also more easily assess the goodness of fit and choose the number of clusters by using information criteria (see below). A second advantage is that with statistical models different aspects of the results can be controlled, the primary of which is the cluster shape. Assumptions on the data generating model mostly correspond to different cluster shapes and similar assumptions, and can be easily interpreted and tested. The drawbacks are that, as always, when we specify a parametric model for the data generating mechanism we may be far from reality and thus obtain biased estimates; and that the strategies for optimization are slightly more cumbersome and slow to convergence with respect to the conventional k-means algorithm.

In the model based clustering framework, data are viewed as arising from a finite mixture of probability distributions, each of which represents a cluster. Model-based clustering has a long history. Surveys of model based clustering approaches can be found in Bock (1996) and Fraley and Raftery (2002). Related issues, such as choosing the number of clusters, are discussed in McLachlan and Basford (1988), Banfield and Raftery (1993), McLachlan and Peel (2000). Approaches specifically tailored for gene expression data can be found in Yeung *et al.* (2001) and McLachlan *et al.* (2002).

In the following, we will assume that data arise from a mixture of Gaussian distributions. Hence, each cluster can be respresented by two parameters: a vector of means $\mu_c \in \mathbb{R}^p$ and a $p \times p$ covariance matrix Σ_c; $c = 1, \ldots, k$. We now outline methods for estimating μ_c and Σ_c, and consequently the class labels matrix U. It will become clear that while the k-means algorithm is based on the Euclidean distance (7.2), model based clustering methods, at least under normality assumptions, are based on the squared Mahalanobis distance

$$d_{ic}^2 = (x_i - \bar{x}_c)^T \hat{\Sigma}_c^{-1} (x_i - \bar{x}_c), \tag{7.4}$$

where as before $\bar{x}_c$ denotes the estimated centroid for the cth cluster, and $\hat{\Sigma}_c$ denotes an estimate for Σ_c. Note that we use the same notation d_{ic}^2 to denote the Euclidean and Mahalanobis distances, given that the former is a special case of the latter where $\hat{\Sigma}_c$ is assumed to be the identity matrix of size p.

7.4.1 Likelihood inference

There are two possibilities for estimating the parameters at stake. One is based on formulating and maximizing the classification likelihood (Celeux and Govaert, 1992, 1995), the other on formulating and maximizing the mixture likelihood (Fraley and Raftery, 1998; Dempster *et al.*, 1977). In both cases, estimates are obtained by implementing an iterative algorithm which alternates two steps: in the first step, maximization is performed conditionally on the current estimates for the parameters; in the second step the parameters are updated conditionally on the first step. We clarify by briefly describing the two approaches in the following.

The classification likelihood can be defined as:

$$\prod_c \prod_i \phi_p(x_i, \mu_c, \Sigma_c)^{u_{ic}}, \tag{7.5}$$

where as before u_{ic} is an indicator of the ith subject belonging to the cth cluster.

The mixture likelihood is based additionally on cluster weights, or cluster *prior probabilities* $\pi_1, \ldots, \pi_k$, collected in the vector π, so that *a priori* $\Pr(u_{ic} = 1) = \pi_c$, $c = 1, \ldots, k$. The mixture likelihood can be expressed as:

$$\prod_i \sum_c \pi_c \phi_p(x_i, \mu_c, \Sigma_c). \tag{7.6}$$

In practical terms, direct maximization of (7.5) or (7.6) is cumbersome and algorithms are easily trapped into local peaks. A solution is given by the EM algorithm. We describe it for the mixture likelihood. The EM algorithm to maximize (7.6) works with the *complete likelihood*

$$\prod_i \prod_c (\pi_c \phi_p(x_i, \mu_c, \Sigma_c))^{u_{ic}}, \tag{7.7}$$

where the unknown u_{ic} is treated as fixed. The equivalent for (7.5) is obtained by assuming $\pi_c = 1/k$, which favors clusters of approximately equal size. In the first case, the classical EM is obtained and the mixture likelihood is maximized. In the second case, the classification EM is obtained and the classification likelihood is maximized.

The substantial difference between the classification and classical EM algorithms is at the E-step. For the classification EM, u_{ic} is estimated as either 0 or 1, where the ith observation is assigned to the most likely cluster conditionally on the current estimates for μ and Σ. Formally, let

$$\tilde{c}_i = \arg \max_c \phi_p(x_i, \mu_c, \Sigma_c),$$

then $u_{i\tilde{c}_i} = 1$ and $u_{ic} = 0$ for $c \neq \tilde{c}_i$. After straightforward algebra, using the explicit expression for the multivariate normal density, it can be seen that this corresponds to assign the ith observation to the closest centroid in terms of Mahalanobis distance d_{ic}^2.

The classical EM instead proceeds by computation of the posterior expectation of (7.7), which is obtained by setting (Dempster *et al.*, 1977)

$$u_{ic} \propto \pi_c \phi_p(x_i, \mu_c, \Sigma_c). \qquad (7.8)$$

At the M-step, π, μ and Σ are estimated conditionally on u_{ic}. There are closed form expressions, which are easily given as

$$\mu_c = \frac{\sum_i u_{ic} x_i}{\sum_i u_{ic}},$$

$$\Sigma_c = \frac{\sum_i u_{ic}(x_i - \mu_c)(x_i - \mu_c)^T}{\sum_i u_{ic}},$$

and, when needed,

$$\pi_c \propto \sum_i u_{ic}.$$

The value at convergence of u_{ic} can be used for assigning each observation to one of the k clusters. With the classification EM this is straightforward as u_{ic} at convergence already gives the most likely classification. With the classical EM, (7.8) can be interpreted as the posterior probability that the ith subject belongs to the cth cluster. An obvious (but not the only available) approach for cluster assignment is then to use a Maximum-A-Posteriori criterion, that is, a final E-step of the classification type for which

$$\tilde{c}_i = \arg\max_c \pi_c \phi_p(x_i, \mu_c, \Sigma_c),$$

and $u_{i\tilde{c}_i} = 1$ and $u_{ic} = 0$ for $c \neq \tilde{c}_i$.

Each iteration of the (classification) EM algorithm is made of one expectation and one maximization step. The algorithm is initialized from an arbitrary candidate solution, and at each iteration the target criterion (7.5) or (7.6) is bound to increase. The number of iterations is usually not fixed in advance, and the algorithm is run until two iterations give approximately the same likelihood, that is, the increase in the objective function from the previous iteration is seen to be below a pre-specified tolerance. To summarize, we report the general iteration of the EM and classification EM in Algorithm 7.2 and 7.3, respectively.

A detailed comparison of the EM and classification EM can be found in Celeux and Govaert (1993). It can be argued that maximizing the classification likelihood is often more appropriate for clustering. The most important difference is that the classification EM converges faster than the classical EM, that is, with a much smaller number of iterations. Furthermore, classification EM reduces to the conventional k-means algorithm when Σ_c is assumed to be diagonal (and $\pi_c = 1/k$). On the other hand, there may be issues with the

---

Algorithm 7.2 EM algorithm
E-step
for $i = 1, \ldots, n$ do

$$u_{ic} = \frac{\pi_c \phi_p(x_i, \mu_c, \Sigma_c)}{\sum_c \pi_c \phi_p(x_i, \mu_c, \Sigma_c)}.$$

end for
M-step
for $c = 1, \ldots, k$ do

$$\mu_c = \frac{\sum_i u_{ic} x_i}{\sum_i u_{ic}},$$

$$\Sigma_c = \frac{\sum_i u_{ic}(x_i - \mu_c)(x_i - \mu_c)^T}{\sum_i u_{ic}},$$

$$\pi_c = \frac{\sum_i u_{ic}}{\sum_i \sum_c u_{ic}},$$

end for

---

properties of the estimates at convergence of the classification EM (Scott and Symons, 1971).

A final issue regards initialization for the parameters. The EM and classification EM algorithm are well known to be dependent on the initial values. A simple and common solution is to repeatedly run the algorithm from a number (say, 20) of different starting values. These values can be chosen by pilot runs, use of simpler models, and random perturbation of previously used initial solutions, or solutions obtained at convergence of the algorithm. It shall be noted that only a set of parameters involved in the E-step *or* in the M-step needs to be initialized. To be more precise, one can proceed by initializing μ, Σ and, if used, π. After initialization of those parameters, an E-step can be performed to obtain u_{it}. Similarly, one can initialize u_{it} and then perform an M step to obtain the corresponding estimates for μ, Σ and π.

7.4.2   Distribution of component densities

The number of clusters and the distribution of the component densities are two key assumptions. Different specifications lead to completely different statistical models and interpretation of the results.

In some cases, the distribution of one or more component densities is taken to be different than the Gaussian. Student's T distributions can be used to deal with observations far from any mode and take into account fat tails, skewed distributions can also be used to permit peculiar cluster shapes. Works in this spirit are Banfield and Raftery (1993), Frühwirth-Schnatter and Pyne (2010).

Algorithm 7.3 Classification EM algorithm

Classification E-step
for $i = 1, \ldots, n$ **do**

$$\tilde{c}_i = \arg\max_c \phi_p(x_i, \mu_c, \Sigma_c),$$

$u_{i\tilde{c}_i} = 1$
for $c \neq \tilde{c}_i$ **do**
    $u_{ic} = 0$
end for
end for
M-step
for $c = 1, \ldots, k$ **do**

$$\mu_c = \frac{\sum_i u_{ic} x_i}{\sum_i u_{ic}},$$

$$\Sigma_c = \frac{\sum_i u_{ic}(x_i - \mu_c)(x_i - \mu_c)^T}{\sum_i u_{ic}},$$

end for

McLachlan and Peel (2000) introduced the use of mixtures of Student's T distributions. The multivariate T is also used in Humburg *et al.* (2008). Fraley and Raftery (1998) considered a finite mixture of multivariate normal distributions with an additional component modeled as a Poisson process to handle noisy data; Hennig (2004) included an additional component distributed like an improper uniform. A recent review can be found in Lee and McLachlan (2013).

In this section we consider how to restrict, rather than make more flexible, the Gaussian mixture. This can be easily accomplished through assumptions on one or more of the population scatter matrices Σ_c, $c = 1, \ldots, k$. A general way of doing this is through the parameterization (Celeux and Govaert, 1995; Fraley and Raftery, 2002):

$$\Sigma_c = |\Sigma_c|^{1/p} V_c \frac{\Lambda_c}{|\Sigma_c|^{1/p}} V_c^T. \tag{7.9}$$

The parameterization (7.9) is based on three elements: $|\Sigma_c|^{1/p}$, V_c and $\Lambda_c/|\Sigma_c|^{1/p}$, where as usual V_c denotes the matrix of eigenvectors and Λ_c the diagonal matrix of eigenvalues of Σ_c. These three elements control the volume, orientation and shape of the cth cluster, respectively.

When no assumptions are formulated, there are arbitrary and cluster specific size and orientation of elliptically contoured clusters. The normality assumption in fact implies that clusters are elliptically contoured. The number of parameters involved is rather large, being of the order of magnitude of kp^2.

TABLE 7.3

Models arising from assumptions on volume, shape and orientation. The three letter codes designate assumptions of equality (E) or varying (V) volume, shape or orientation, respectively. A letter I indicates a spherical shape or an axis-aligned orientation.

Code	Description
EII	spherical, equal volume
VII	spherical, varying volume
EEI	diagonal, equal volume and shape
VEI	diagonal, equal shape
EVI	diagonal, equal volume, varying shape
VVI	diagonal, varying volume and shape
EEE	ellipsoidal, equal volume, shape and orientation
VEE	ellipsoidal, equal shape and orientation
EVE	ellipsoidal, equal volume and orientation
VVE	ellipsoidal, equal orientation
EEV	ellipsoidal, equal volume and shape
VEV	ellipsoidal, equal shape
EVV	ellipsoidal, equal volume
VVV	ellipsoidal, varying volume, shape, and orientation

By formulating assumptions, different clustering models are obtained. The number of free parameters can be reduced substantially. Assumptions are formulated so that certain, or even all, characteristics are equal across clusters. For instance, it can be assumed that $|\Sigma_c|^{1/p}$ is constant over c, therefore obtaining clusters of the same volume. Another family of assumptions regards V_c. It can for instance be assumed that clusters all have the same orientation, or that Σ_c is diagonal. Finally, it can be assumed that all elements of Λ_c are equal, therefore having *spherical* clusters. Fourteen models arising from (7.9) are discussed in Celeux and Govaert (1995). A unified estimation and selection procedure is currently not available. The common approach is to fit a selection of possible models, or even all of them, and use the one corresponding to the minimizer of the Bayesian information criterion (BIC) (Schwarz, 1978). BIC is obtained summing minus twice the log-likelihood with the number of parameters multiplied by the logarithm of the sample size. Consequently, for fixed number of parameters, models with larger log-likelihood will be preferred. Models with a large number of parameters are penalized, hence the increase in goodness of fit (i.e., log-likelihood) when increasing model complexity must exceed the loss in simplicity (i.e., the increased number of parameters). The requirement is more and more stringent as the (log) sample size grows.

A library named `mclust` is provided in `R` which allows to estimate different models and choose the best set of assumptions. The set of possible models is summarized in Table 7.3. For a complete description of `mclust` refer to Fraley and Raftery (1999, 2003).

It can be noted that the EM Algorithm 7.2 can be employed regardless of the assumptions formulated on (7.9). The only difference is at the M-step, where maximization must be performed conditionally on constraints. This leads to different expressions for the estimates of Σ_c, $c = 1, \ldots, k$; these expressions are often available in closed form and involve either pooling estimates across clusters, or separately pooling one or more elements of the parameterization (7.9).

We conclude by noting that some well-known clustering algorithms correspond at least approximately to certain probability models obtained under assumptions on (7.9). The conventional k-means algorithm, for example, is equivalent to the classification EM under the conditions that

$$|\Sigma_c| = |\Sigma_d|,$$

for all $c, d = 1, \ldots, k$ and that $\Sigma_c / |\Sigma_c|^{1/p}$ is the identity matrix; that is, under the assumption that $\Sigma_c = \sigma^2 I_p$ for all $c = 1, \ldots, k$; where I_p denotes the identity matrix of dimension p. Other popular constraints include the homoscedastic case in which $\Sigma_c = \Sigma_d$.

7.4.3   Examples of model-based clustering

We give a brief illustration of model-based clustering algorithms, interpretation of their results, and a hint towards lack of robustness in the following examples.

We briefly revisit the G8, metallic oxide and water treatment plant data examples.

We fit Gaussian mixture models using function `Mclust` in library `mclust`.

As far as the G8 data are concerned, the best model selected is EII with $k = 3$. We use BIC both to choose the model structure and the number of groups (which is the default with function `Mclust`). The model structure chosen is the simplest possible and corresponds to k-means. This outcome is not surprising as the sample size is so low that it is difficult to justify more complex models. The final cluster labels are given in Table 7.4. When these are compared with labels in Table 7.1, it can be seen that model based clustering grants a much more coherent result in terms of clustering. This result involves identifying Spain and Italy as two separate and quite different outliers, and the other countries as a unique cluster. This clustering is meaningful and coherent, albeit useless as we do not actually manage to discriminate among non-outlying countries.

Let us now consider the metallic oxide data. For these data we select model EEE, and not unsurprisingly we obtain exactly the same cluster labels as k-means both when $k = 2$ and $k = 3$. Here we fixed the number of clusters. As a consequence, even allowing for elliptical clusters, which is slightly more flexible than the spherical cluster assumption behind k-means, we still obtain exactly the same results. We are not able to predict the type of raw material, and the Rand index is always very low.

TABLE 7.4

G8 macroeconomic data. Cluster labels using `mclust` with $k = 3$. Results are based on standardized measurements.

Country	Labels
FRA	1
GER	1
GBR	1
ITA	2
SPA	3
USA	1
JAP	1
CAN	1

We now turn to the water treatment plant example. The best model selected by `mclust` is once again EEE when we fix $k = 3$, but we obtain a substantially different clustering with respect to k-means, where the number of observations in each cluster is 25, 345 and 10. It shall be here remarked that model based clustering methods allow for slightly more unbalanced clusters. In presence of outliers this results basically in few clusters made of outliers and a single cluster with the bulk of the observations, as with the G8 example.

When we look at the reduced data in Table 7.5 we can see that the first and third cluster are made of outlying days with very heavy load at input, and are separated by the average measurements at the second settler and output. There is a marked increase in the average measurements for the third cluster when passing from the second settler to the output. This is rather surprising and may be due to the presence of contamination. Finally, we note that the bulk of the data shows means that are very close to zero, which is the global mean after standardization. This is not very informative and once again a by-product of the heavily unbalanced clusters obtained.

TABLE 7.5

Water treatment plant data. Average of standardized measurements by location and cluster as identified by `mclust` when $k = 3$.

	Cluster 1	Cluster 2	Cluster 3
Input	0.20	-0.03	0.43
First Settler	0.20	-0.04	0.88
Second Settler	0.56	-0.04	0.02
Output	0.23	-0.04	0.94

7.5   Choosing the number of clusters

In this section we give some guidelines on how to choose the number of clusters. This corresponds to choose the parameter k for the k-means algorithm. As far as model based clustering is concerned, it is also generally believed that the number of clusters and the number of components k are in a one-to-one correspondence. Given the expected cluster shape, this is true if and only if a k component mixture has got k modes. As a matter of fact, the reader must keep in mind that it may not be the case in some situations, especially when p is large, and even when $p = 2$ the one-to-one correspondence between the number of mixture components and the number of modes may be lost. For more details, see Ray and Lindsay (2005). In such pathological cases, one can still interpret the model as reducing the sample size to k, but at the price of having one or more *multimodal* clusters. See also Carreira-Perpinan and Williams (2003).

Choosing the number of clusters is still an open problem in cluster analysis. It is so in our opinion because the optimal number of clusters is also heavily dependent on the cluster shape, where a smaller number of elliptical, with respect to spherical, clusters may be enough to span a data set. Similar reasoning can be applied to any family within the general decomposition (7.9), and the practitioner should keep in mind that a comparison of the number of clusters selected under different assumptions may not be sensible. While an optimal cluster shape can often be selected, it leads to different interpretations. A better route is that the practitioner specifies cluster shapes according to the desired definition of group.

As a further preliminary note, in confirmatory cluster analysis the number of clusters is actually known in advance. Furthermore, as pointed out by Climer and Zhang (2006), it often occurs that only a limited range of values are of interest.

A first approach to choosing the number of clusters is given by proceeding manually, by comparing cluster profiles for different values of k. A location and scatter estimate can be provided for each centroid, in order to visually evaluate the separation and tightness of the clusters. A simple general idea is that too many clusters have been specified if the centroids are not well separated, while the sample has been reduced too much if variability within some cluster is too large. To evaluate the first issue, one can simply compare location estimators across clusters and check if these are far apart. If two or more clusters have centroids that are close to each other, then k should be decreased. To evaluate the second issue one can look separately at each scatter estimator. If at least one cluster shows a large variability, it may be that observations that should belong to different groups have been merged. In this case, k should be increased.

Manual evaluation of k may be informative but cumbersome, especially when p is large. There are appropriate tools that are designed to evaluate separation and tightness of each cluster, and help in tuning k. One of the most popular suggestions is given in Rousseeuw (1987) and detailed in Chapter 2 of Kaufman and Rousseeuw (1990). It is known as the Silhouette statistic s_i. A quality of assignment of each observation $i = 1, \ldots, n$ to its cluster is summarized by $s_i = 0$ if i is a singleton, and otherwise

$$s_i = \frac{b_i - a_i}{\max(a_i, b_i)},$$

where a_i is the average dissimilarity between the ith observations and all other points in the cluster; b_i is the smallest average dissimilarity with the other clusters, that is, the average dissimilarity with all points in the cluster closest to the ith observation. The average Silhouette

$$\bar{s} = \frac{1}{n} \sum_i s_i$$

is then used to summarize the quality of clustering based on k groups; and the final k can be set as the maximizer of $\bar{s}$.

A more recent proposal is the GAP statistic of Tibshirani *et al.* (2001), which compares the expected and observed mean square error obtained from sample reduction, and looks for the largest gap between two consecutive values of k. The number of groups can be simply set as the largest GAP statistic. This has been experienced to lead to overly complex models, as even natural clusters may be split. A refined method which we will use later is proposed in Tibshirani *et al.* (2001), taking also into account the uncertainty in estimating the GAP. The standard errors are evaluated using a simple bootstrap procedure, and the chosen k is the smallest such that the corresponding GAP statistic exceeds the one based on $k + 1$, minus its standard error. Formally, let $GAP(k)$ denote the GAP statistic associated with k clusters and $se_{GAP}(k)$ its standard error. The number of clusters is set as the smallest k for which

$$GAP(k) \geq GAP(k + 1) - se_{GAP}(k + 1),$$

and $k = 1$ if the condition above is always false. There are many other measures to evaluate clustering and choose k. Important references are Milligan and Cooper (1985) and Gordon (1999). See also Hartigan (1978) for a more general discussion.

In model based clustering, goodness of fit criteria can be obtained as functions of the likelihood at convergence. Popular approaches are the so called *information criteria*, which are based on penalizing the value of the likelihood at convergence with a function of the number of free parameters. Different models can be then compared and the one corresponding to the best (penalized) fit used for inference. One of the first approaches is the Akaike information criterion (AIC) of Akaike (1973), and another is the Bayesian information

TABLE 7.6
Metallic oxide data. Centroid estimates and within-group standard deviations for 4-means. In parentheses, the number of observations in each cluster.

Cluster	Cluster Means							
1 (10)	3.33	3.17	3.28	3.32	3.52	3.48	3.36	3.28
2 (18)	4.01	3.97	3.94	3.96	3.96	3.99	4.02	4.03
3 (1)	3.30	3.40	3.90	4.00	2.20	2.30	2.40	2.70
4 (2)	0.55	0.65	0.55	0.65	0.65	0.75	0.45	0.00
	Cluster Standard Deviations							
1 (10)	0.21	0.44	0.26	0.35	0.49	0.37	0.35	0.37
2 (18)	0.30	0.35	0.25	0.30	0.31	0.28	0.19	0.25
3 (1)	0	0	0	0	0	0	0	0
4 (2)	0.49	0.07	0.35	0.07	0.49	0.49	0.35	1.41

criterion (BIC) (Schwarz, 1978), which has a Bayesian interpretation and is known to be consistent under certain conditions. On the other hand, AIC may be more appropriate when the sample size is small and provides optimal results from a predictive point of view. There are other possibilities (e.g., the cAIC; see Burnham and Anderson (2002)). A particularly interesting alternative is given by the ICL criterion (Biernacki *et al.*, 2000), which may be relevant specifically for estimating the number of modes rather than the number of mixture components.

The methods discussed in this section are inherently non-robust. Extension of the average Silhouette and other statistics to the robust setting is often straightforward (e.g., a trimmed silhouette may be based only on clean observations as detected by any robust method). A formally robust proposal for choosing the number of groups is given by the trimmed likelihood curves of García-Escudero *et al.* (2011), which will be discussed later.

We conclude this section with a simple example.

7.4 Example (Selection of k for the metallic oxide data) *We revisit here the metallic oxide data, and pretend the number of groups is unknown.*

We begin working manually. The procedure is equivalent regardless of the clustering method, hence we develop it only for k-means. First of all, we set an initial number of clusters which could be slightly overly estimating the true k. For the metallic oxide example, this could be $k = 4$. We then run 4-means and produce a table with means and standard deviations for each cluster. These can be found in Table 7.6. We see that when $k = 4$ the first and second group are not well separated, as no difference between the $p = 8$ means is markedly large as compared to the standard deviations. This implies that we should decrease the number of groups to $k = 3$.

For reasons of space we do not repeat this reasoning, which would lead to the same conclusions, when $k = 3$. We show the centroids and standard deviations for $k = 2$ and $k = 1$ in Table 7.7.

TABLE 7.7
Metallic oxide data. Centroid estimates and within-group standard deviations
for when $k = 2$ and $k = 1$ based on k-means. In parentheses, the number of
observations in each cluster.

Cluster								
	\multicolumn{8}{c}{Cluster Means}							
1	0.55	0.65	0.55	0.65	0.65	0.75	0.45	0.00
2	3.75	3.68	3.71	3.74	3.75	3.76	3.73	3.72
	\multicolumn{8}{c}{Cluster Standard Deviations}							
1	0.49	0.07	0.35	0.07	0.49	0.49	0.35	1.41
2	0.43	0.54	0.40	0.44	0.52	0.48	0.48	0.50
	\multicolumn{8}{c}{Cluster Means}							
1	3.55	3.48	3.51	3.54	3.55	3.56	3.52	3.48
	\multicolumn{8}{c}{Cluster Standard Deviations}							
1	0.91	0.92	0.88	0.88	0.93	0.89	0.94	1.08

TABLE 7.8
Metallic oxide data. Silhouette (SIL), GAP and related standard error
($\text{se}_{GAP}(k)$).

k	Sil	$GAP(k)$	$\text{se}_{GAP}(k)$
1		0.47	0.07
2	0.80	0.39	0.05
3	0.43	0.46	0.05
4	0.37	0.43	0.05
5	0.23	0.42	0.05
6	0.25	0.41	0.05
7	0.21	0.40	0.05
8	0.23	0.38	0.05
9	0.19	0.38	0.05

It can be seen that when $k = 2$ the two groups are very well separated.
Most standard deviations are rather small as compared to the estimated mean,
hence the within group variability may not be too large (otherwise we would
have increased k by one unity). When $k = 1$ we only look at the standard
deviations. These are not alarmingly large, hence also $k = 1$ would be acceptable. Nevertheless, the standard deviations are strongly inflated in passing
from $k = 2$ to $k = 1$, hence we would recommend $k = 2$.

We now use automatic selection criteria. With k-means we compute the
Silhouette and GAP statistics. The standard errors for the GAP statistic are
estimated using 1000 bootstrap repetitions. These are reported in Table 7.8.
With model based clustering we compare BIC values for different possible
models. These are reported in Table 7.9, where an empty cell indicates that
the model is not feasible.

TABLE 7.9
Metallic oxide data. BIC for model-based clustering based on different number of clusters and different assumptions.

k	EII	VII	EEI	VEI	EVI	VVI	
1	-690.3	-690.3	-712.3	-712.3	-712.3	-712.3	-277.4
2	-399.8	-403.1	-418.3	-421.0	-433.5	-426.4	**-261.9**
3	-312.0	-299.8	-323.1	-314.7	-336.3	-328.3	-270.8
4	-301.8	-302.3	-317.5	-317.6			-289.8
5	-319.4	-312.7	-329.4	-322.0			-301.3
6	-312.9		-320.2				-323.4
7	-298.0		-303.5				-313.3
8	-301.2		-312.5				-326.5
9	-310.7		-327.1				-348.3
k	EEV	VEV	VVV				
1	-277.4	-277.4	-277.4				
2	-332.9	-316.7					
3	-392.2	-388.9					
4	-480.6	-439.2					
5	-397.1	-376.4					
6	-518.0						
7	-635.6						
8	-723.2						

It can be seen that the largest Silhouette is obtained when $k = 2$. The optimal model based clustering formulation as suggested by the BIC criterion is only slightly more complex, being the ellipsoidal model (EEE) with $k = 2$. As far as the GAP statistics are concerned, the largest GAP is obtained with $k = 1$, hence the GAP statistic would lead to set $k = 1$. Also the refined method in Tibshirani et al. (2001), which also takes into account $se_{GAP}(k)$, leads to $k = 1$.

7.5 R Illustration (Clustering metallic oxide data) For the metallic oxide example, data are stored in a data frame `fell`. Model based and k-means clustering, and their agreement with the true labels, were obtained with the code

```
> source("rand.index.r")
> cl2=kmeans(fell[,-1],2)$cl
> cl3=kmeans(fell[,-1],3)$cl
> rand.index(unclass(fell[,1]),cl2)
> # rand index after discarding the two outliers
> rand.index(unclass(fell[-which(cl3==3),1]),
+ cl3[cl3!=3])
> library(mclust)
> cl2=Mclust(fell[,-1],2)
```

```
> cl3=Mclust(fell[,-1],3)
> rand.index(unclass(fell[,1]),cl2)
> # rand index after discarding the two outliers
> rand.index(unclass(fell[-which(cl3==3),1]),
+ cl3[cl3!=3])
```

For choosing the number of clusters we instead proceeded heuristically at first as follows

```
> km4=kmeans(fell[,-1],4)
> km4$centers
> apply(fell[,-1],2,function(x) tapply(x,km4$cl,sd))
> km3=kmeans(fell[,-1],3)
> km3$centers
> apply(fell[,-1],2,function(x) tapply(x,km3$cl,sd))
> km2=kmeans(fell[,-1],2)
> km2$centers
> apply(fell[,-1],2,function(x) tapply(x,km2$cl,sd))
```

and then used formal methods as in

```
> library(fpc)
> sil=rep(NA,9)
> for(k in 2:9) {
+ sil[k]=cluster.stats(dist(fell[,-1]),
+ kmeans(fell[,-1],centers=k)$cl)$avg.silwidth}
> which.max(sil)
> gap=clusGap(fell[,-1],kmeans,K.max=9,nstart=10,
+ iter.max=100,B=1000)
> print(gap,method="Tibs2001")
> cbind(sil,gap$Tab[,3:4],Mclust(fell[,-1])$BIC)
```

8

Robust Clustering

In this chapter we will introduce and motivate robust unsupervised sample reduction methods which are not based on probabilistic assumptions on the data generating distribution. We will start from one of the first attempts to robustify sample reduction, namely, partitioning around medoids. We then will proceed with trimmed k-means. Finally, we review snipped k-means, a generalization of trimmed k-means tailored for component-wise outliers. The reader is also referred to Banerjee and Dave (2012) for a review of some topics which are not covered here, including robust hierarchical clustering and robust fuzzy clustering.

We give a short general motivation of robust clustering. It may be argued that outliers could be isolated by simply increasing k. Nevertheless, the number of clusters must be set in advance, and we could have not taken into account the presence of outliers. Most importantly, stability of k-means may be seriously affected by outliers even for large k. Finally, the output of small clusters, or even singletons, may lead to problems in interpretation and use of the clustering for future prediction, classification and/or resource allocation. Another preliminary comment in regards to the possibility to perform outlier detection before sample reduction. This is rather common but it must be kept in mind that any procedure performing outlier detection before the actual analysis might be severely prone to masking. Outliers are unusual observations *within* their cluster, while they could be perfectly acceptable if considered with respect to the entire data set. This is particularly true when considering bridge points, but not only. Given that we do not know clusters *or* outliers in advance, the only possibility is then to deal with outliers and clustering *simultaneously*.

8.1 Partitioning Around Medoids

Partitioning Around Medoids (PAM) is proposed in Kaufman and Rousseeuw (1987) and described in full detail in Chapter 2 of Kaufman and Rousseeuw (1990). It proceeds by selecting k observations in the data set, the *representative objects* or *medoids*, and then assigning the remaining $n - k$ observations to the cluster whose representative object is the closest. The key idea is that centroids are not computed as a location estimator of all observations assigned to a cluster, but they correspond to exactly one of the observations. The optimal representative objects are those minimizing the sum of the distances between them and all other units in the cluster. More formally, using the Euclidean distance our objective function for PAM is

$$\sum_{i=1}^{n} \min_{c=1,\ldots,k} \sum_{j} (x_{ij} - x_{cj})^2.$$

The former expression is quite similar to that of k-means, with the only difference that the arbitrary centroids of k-means $\bar{x}_c$, $c = 1, \ldots, k$ must now be chosen within the observed data x_i, $i = 1, \ldots, n$. Even with this substantial difference with k-means, PAM leads to approximately spherical clusters. Note that in the objective function we still use a quadratic distance. Similar results can be obtained by using the so-called k-medoid method, where the $\mathcal{L}_1$ rather than $\mathcal{L}_2$ norm is used. See for instance Vinod (1969).

Kaufman and Rousseeuw (1990) report that PAM is more robust with respect to outliers than k-means. This is reasonable as one, or even few, isolated aberrant observations will unlikely be selected as a representative object. These observations will be far from each of the k bulks of data. The resulting medoids, not being based on summary statistics like the sample mean, will be unaffected by these outliers. Despite this seemingly encouraging reasoning, it is demonstrated in García-Escudero and Gordaliza (1999) that PAM is *not* formally robust. As a matter of fact, a single outlier may break down the method, which therefore is not globally robust.

PAM is especially recommended if one is interested in the representative objects themselves, which may be very useful for data reduction or characterization purposes. There may be various reasons for this. An intriguing example is given in Kaufman and Rousseeuw (1990), where objects are areas within a city, and the problem of clustering is linked with finding k locations for placement of power plants. In a situation like this, non-existing centroids obtained by averaging are almost useless; while the representative objects are exactly the optimal places for building new power plants (if one wants to minimize the distance between the clients and the k servers). The representative objects might be selected for further investigation in some applications. The k representative subjects can be contacted for information which would be too expensive to collect on the entire sample, for instance. It can be generally believed that the discoveries on each representative subject can be extended to the other subjects in the same cluster. These can be in fact expected to be similar to their representative object, at least with respect to the characteristics originally measured.

A general formulation for the objective function of PAM is given by

$$\sum_{i=1}^{n} \min_{c=1,\ldots,k} d(x_i, x_c). \tag{8.1}$$

Algorithm 8.1 The *build* phase of the PAM algorithm

Select the first representative object
Let $c_1 = \arg\min_j \sum_i d(x_i, x_j)$.
Select the other $k-1$ representative objects
for $h = 2, \ldots, k$ **do**

$$c_h = \arg\max_i \sum_{j=1}^{n} \max(\min_{l=1,\ldots,h-1} d(x_j, x_{c_l}) - d(x_i, x_j), 0) \qquad (8.2)$$

end for
Let $c_1, \ldots, c_k$ denote the indices of the k initial medoids.

Given that k units x_c, $c = 1, \ldots, k$ can be chosen so that (8.1) is minimized *regardless* of the distance function used $d(\cdot, \cdot)$, PAM does not only deal with interval-scaled measurements. Furthermore, in order to use PAM we do not need to have a record of the raw data, but can deal directly with dissimilarities $d(\cdot, \cdot)$.

In order to minimize (8.1), an iterative algorithm is used. The algorithm can be schematically divided in two phases. In the first, commonly known as the *build* phase, the initial representative objects are chosen. The *build* phase starts by selecting the most central observation, that is, the one minimizing the sum of the dissimilarities. Then, other $k-1$ steps are performed iteratively including in the set of initial representative objects the observation decreasing the objective function as much as possible. To this end, the gain in using the ith observation as a medoid is expressed as the difference between the total distance of the other observations to their closest medoid, and the distance of the other observations to the ith one, when this is smaller. A formal expression is given in (8.2).

At the end of the *build* phase, there is a *swap* phase. In the *swap* phase, the medoids and ordinary observations are switched one at a time until the objective function can not be improved anymore. This leads to a local minimum for (8.1). It can also be noted that given that all potential swaps are iteratively considered, when there are no tied distances the results do not depend on the ordering of the objects.

The two phases are separately described in Algorithm 8.1 and 8.2. We note that the usual formulation of Algorithm 8.2 is slightly more complex, but that the one we give is absolutely equivalent and with the same computational complexity of the original proposal in Kaufman and Rousseeuw (1990).

A variation of PAM for large data sets, CLARA, is proposed in Chapter 3 of Kaufman and Rousseeuw (1990).

Algorithm 8.2 The *swap* phase of the PAM algorithm

$flag = TRUE$
while $flag$ **do**
  Current loss
  Let $l_{old} = \sum_{j=1}^{n} \min_{l=1,\ldots,k} d(x_j, x_{c_l})$
  If no improvement, quit the algorithm
  $flag = FALSE$
  for $h = 1, \ldots, k$ **do**
    for $i \neq \{c_1, \ldots, c_k\}$ **do**
      Compute the loss if swapping c_h **with** i
      Let $l_{new} = \sum_{j=1}^{n} \min(\min_{l \neq h} d(x_j, x_{c_l}), d(x_j, x_i))$
      if $l_{old} > l_{new}$ **then**
        An improvement is seen with the swap
        $c_h = i$
        $l_{old} = l_{new}$
        The algorithm is run for an additional iteration
        $flag = TRUE$
      end if
    end for
  end for
end while

8.2   Trimmed k-means

Trimming is a simple and intuitive method for robustification, being based on discarding a (usually fixed) proportion of observations. These shall be selected as the most extreme, that is, the most distant from the centroid. When $p = k = 1$ this is straightforward. In all cases, selection of the observations that must be trimmed and estimation shall be done simultaneously (Gordaliza, 1991).

We can summarize trimmed robust clustering as follows: a (maximum) proportion of ε observations is allowed to be almost arbitrarily contaminated, and therefore discarded. The remaining observations are clean and can be used to estimate the k centroids. As a consequence, two parameters shall be specified: one is k, the number of groups, the second is ε, the proportion of contaminated observations. Guidelines for fixing k can be found in Section 7.5, while guidelines for fixing ε in Section 8.4 below.

The idea behind trimming is very simple, but estimation may be complicated by the fact that it may not be possible to detect contaminated observations, which may be bridge points for instance. Trimming approaches usually proceed by iteration of trimmed least squares estimation of centroids and so called *concentration steps* to flag outliers with respect to the current estimates (e.g., Rousseeuw and Van Driessen (1999), Gallegos and Ritter (2005), Far-comeni (2009b)). For more details see also Rousseeuw (1984) and Rousseeuw and Van Driessen (2006). Concentration steps are based on identification of the $n\varepsilon$ largest Mahalanobis distances, and trimming of the corresponding observations, where the distances are computed with respect to the closest centroid.

8.2.1 The double minimization problem involved with trimmed k-means

The optimization problem connected with trimmed k-means can be expressed as a double minimization problem as follows:

$$\min_{z \in \mathcal{Z}} \min_{\bar{X}} \sum_{i=1}^{n} z_i \min_{c=1,\ldots,k} \sum_{j=1}^{p} (x_{ij} - \bar{x}_{cj})^2, \qquad (8.3)$$

where z denotes a binary vector of length n, and $\mathcal{Z}$ denotes the set of all possible binary vectors whose sum is the closest integer to $n(1 - \varepsilon)$. Notice that this definition includes the conventional k-means as a limiting case when $\varepsilon = 0$. The objective function (8.3) is very similar to the conventional k-means objective function (7.3), with the difference that observations for which $z_i = 0$ do not contribute to the loss, as distances are multiplied by z_i in (8.3).

Minimization of the loss function above proceeds iteratively by alternating a concentration step and least squares estimation based on the observations not flagged for trimming, as outlined in Algorithm 8.3.

The computational complexity of the algorithm is $O(\max(nkp, n\log(n)))$ per iteration, due to the sorting involved at the concentration step. There are different aspects of Algorithm 8.3 that shall be here underlined. It shall be noted first of all that trimmed observations are specific to each iteration. The observations are trimmed during the concentration step, based on the $n\varepsilon$ largest Mahalanobis distances from the closest centroid of each observation. Trimmed observations are then not assigned to any cluster, that is, the entire row u_i corresponding to a trimmed observation is set to zero. Due to this fact, the estimate of the centroid matrix does not take into account the values of the trimmed observations. The estimated centroid is therefore not affected by outliers when those are trimmed. The algorithm is iterated until convergence, and repeatedly started from randomly chosen initial solutions as with the conventional k-means algorithm. The final U obtained at convergence simultaneously identifies the optimal cluster for each observation that is not trimmed, and trimmed observations which correspond to a constant row of zeros.

Trimmed k-means estimates always exist and are consistent to the population version of (8.3). They also converge in distribution. It shall be mentioned that trimmed k-means solutions are shown to exist at population level even without any moment conditions. See Cuesta-Albertos *et al.* (1997) and García-Escudero *et al.* (1999).

Algorithm 8.3 Trimmed k-means algorithm

Compute the Mahalanobis distances
for $i = 1, \ldots, n$ **do**

Let $d_{ic}^2 = \sum_{j=1}^{p} (x_{ij} - \bar{x}_{cj})^2$, $c = 1, \ldots, k$.

Let $r_i = \mathrm{argmin}_c d_{ic}^2$.

end for
Concentration step
Sort the distances to the closest centroid as

$$d_{i_1 r_{i_1}}^2 \leq d_{i_2 r_{i_2}}^2 \leq \cdots \leq d_{i_n r_{i_n}}^2$$

Set $z_{i_1} = z_{i_2} = \cdots = z_{i_{n(1-\varepsilon)}} = 1$, all other elements of z are set to zero.
Update the row memberships
for $i = 1, \ldots, n$ **do**

  if $z_i = 1$ **then**

    Set $u_{ir_i} = 1$. All the other elements of the ith row of U are set to 0.

  else

    $u_i = 0$.

  end if
end for
Estimate the centroid matrix
$\bar{X} = (U^T U)^{-1} U^T X$.

The robustness properties of trimmed k-means were firstly studied in García-Escudero and Gordaliza (1999). As with any clustering procedure, also the trimmed k-means has got an infinitesimal universal breakdown point. Ruwet *et al.* (2013) study in detail the restricted breakdown point, first of all formally identifying a class of well clustered data sets for restriction. For data sets within that class, it is shown that the individual breakdown point $\varepsilon^{(i)}$ of trimmed k-means is larger than ε. Hence trimmed k-means can in many cases bear a number of outliers equal to the number of trimmed observations. Farcomeni (2014b) showed furthermore that for trimmed k-means the restricted cell breakdown point is such that

$$\varepsilon^{(c)} \geq \frac{\varepsilon}{p}. \tag{8.4}$$

A consequence is that the number of component-wise outliers that trimmed k-means can bear is only a fraction $1/p$ of the number of trimmed observations. More details on this will be given in the next section.

Other robustness properties of trimmed k-means outlined in García-Escudero and Gordaliza (1999) are that the influence function of trimmed k-means is bounded and has finite gross error sensitivity.

TABLE 8.1
Proportion of contaminated observations under independent component-wise contamination, for different values of contamination probability ε and dimensionality p.

p	5	10	20	50	80	100
$\varepsilon = 0.005$	0.02	0.05	0.10	0.22	0.33	0.39
$\varepsilon = 0.01$	0.05	0.10	0.18	0.39	0.55	0.63
$\varepsilon = 0.05$	0.23	0.40	0.64	0.92	0.98	0.99
$\varepsilon = 0.1$	0.41	0.65	0.88	0.99	1.00	1.00

8.3 Snipped k-means

The basis of this section is that most robust clustering methods, including trimmed k-means, are inherently designed for the Tukey-Huber contamination model. Recall that this contamination model implies structural contamination, that is, some observations are entirely and arbitrarily contaminated. As shown below, these methods may break down under component-wise contamination. Recall that component-wise contamination arises anytime each dimension of the data matrix can be separately contaminated. As a result, most or even all of the observations may be contaminated, even if only slightly. To see this, consider Table 8.1 where we report on the expected proportion of contaminated observations under component-wise contamination, assuming that each entry is independently contaminated with probability ε.

Given that trimming approaches exclude contaminated observations entirely, it may become impossible to work with component-wise contaminated data sets.

An important consideration is that, even if most of the observations may have at least one contaminated entry, under this contamination model only very few of them will be entirely corrupted. In the setting of Table 8.1, when $p = 20$ and $\varepsilon = 0.1$ about 85% of the contaminated observations will have only one or two contaminated entries. These 18 or 19 values shall be retained and used for clustering, resulting in a mild loss of information. In this spirit, Farcomeni (2014b) and Farcomeni (2014a) proposed the idea of snipping, which we outline in the next section.

A full account of component-wise contamination can be found in Alqallaf *et al.* (2009). It shall be here underlined that component-wise contamination models are not scale invariant. Any linear transformation propagates outliers and possibly contaminates additional clean entries. The user shall be aware of the fact that the results of snipping procedures, as the ones outlined in this section, are not invariant under linear transformations of the original data.

We conclude by noting that component-wise contamination is particularly relevant in moderate to high dimension. When p is (relatively) large, departures from model assumptions can be quite more complex than in small dimensional settings. This is seen for instance considering that in small (e.g., $p = 2$) dimensions there may be no practical difference between a structural and a component-wise outlier. See Farcomeni (2015) for a slightly more detailed account on this point.

8.3.1   Snipping   and   the   component-wise   contamination model

The concept of *snipping* is as follows: an observation is snipped when one or more of its dimensions are excluded from the analysis, but at least one is not excluded. Snipped observations can be assigned to clusters, and potentially all observations can be snipped. Snipping is more flexible than trimming as removal of all entries of an observation, which is possible, corresponds to trimming.

This is also useful for the sake of identifying outliers, in that we will be able to explain which dimensions contribute to classify an object as outlying, hence enhancing interpretability.

We now formalize and slightly generalize component-wise contamination. We in fact assume that $np\varepsilon$ entries of the data matrix X are contaminated, for some $\varepsilon > 0$, therefore obtaining entry-wise contamination. Our general contamination model is obtained as a component-wise mixture of a contaminating and a clean density. More formally, let $z_i = (z_{i1}, \ldots, z_{ip})$ denote a binary vector such that $\sum_{ij} z_{ij}$ is equal to the closest integer to $np(1 - \varepsilon)$. The indicator z_{ij} identifies whether the $(ij)^{th}$ entry of the data matrix is free of contamination. When $z_{ij} = 0$, the ith observation has been contaminated in its jth measurement. We assume

$$f(X_i|c) \sim z_i \phi_p(X_i, \mu_c, \sigma^2 I_p) + (1 - z_i)g_i(X_i), \qquad (8.5)$$

for $i = 1, \ldots, n$; where $g_i(\cdot)$ denotes an (almost arbitrary) density in $\mathbb{R}^p$. Model (8.5) brings about component-wise contamination as entries of X_i corresponding to a zero z_i will be generated by the spurious outlier model $g_i(X_i)$. With $f(X_i|c)$ we denote the density of X_i conditional on X_i belonging to the cth cluster, $c = 1, \ldots, k$. Note that when $z_{ij} = 1$ for all $j = 1, \ldots, p$, the cluster assignment is irrelevant as the observation is completely spoiled and is therefore a structural outlier. For more details on (8.5) refer to Farcomeni (2014b). It can be said that (8.5) includes both component-wise, structural and even more complex patterns of contamination. It shall also be noted that the number of clean observations, unlike the Tukey–Huber contamination model, is not fixed when ε is fixed. As a matter of fact, the maximum number of completely spoiled observations is $n\varepsilon$, and the minimum is zero. On the other hand, a minimum of $n\varepsilon$ and a maximum of $\min(np\varepsilon, n)$ observations are contaminated, meaning that at least one entry is generated by a spurious outlier distribution.

We assume that z_i is not random, that is, the data matrix is partitioned into contaminated and clean entries. This is in parallel with the clustering partition, which is not random as well. This is related but slightly different from the partially spoiled contamination model in the definition of Alqallaf *et al.* (2009).

A final comment on (8.5) regards the contamination generating density $g_i(\cdot)$. Let us underline that each outlier is sampled from its own distribution, rather than from a common density $g(\cdot)$. Some assumptions are needed in order to make it possible to estimate μ (and σ^2). These are outlined in Gallegos and Ritter (2005), and generalized to the case of component-wise contamination in Farcomeni (2014a). In practice, the assumptions on $g_i(\cdot)$ imply that outliers are affine independent and that nonregular observations can be ignored when estimating μ. The latter is the consequence of the formal assumption that in any optimal partition with $np\varepsilon$ snipped entries, the non-regular observations could be also obtained by maximizing the likelihood of $g_i(\cdot)$ computed with the same contraints. For more details see also García-Escudero *et al.* (2008) and Ruwet *et al.* (2013).

Given the assumptions, the *snipped k-means* estimates of the centroid μ can be obtained by solving

$$\min_{Z \in \mathcal{Z}} \inf_{\{\mu_1,\dots,\mu_k\} \in \mathbb{R}^p} \sum_{i=1}^{n} \min_{1 \le c \le k} \sum_{j=1}^{p} z_{ij}(x_{ij} - \mu_{cj})^2, \qquad (8.6)$$

where Z is a binary matrix and $\mathcal{Z}$ denotes the set of all subsets of $\mathcal{M}_{\{0,1\}}(n,p)$, the set of binary matrices with n rows and p columns, with exactly $\lceil np\varepsilon \rceil$ entries equal to zero. A zero in Z identifies a discarded entry, no matter if it is discarded because of snipping or because of trimming of its corresponding row. An entire row of zeros identifies a trimmed subject. Note that with trimmed k-means at level ε exactly the same number of entries are removed as with snipped k-means at level ε. The only difference is that with trimming the entries are all aligned in rows, while with snipping they can be anywhere in the data matrix.

8.3.2 Minimization of the loss function for snipped k-means

The problem of minimization of the loss (8.6) could be tackled as with the trimmed k-means problem by alternating concentration steps. It is noted in Farcomeni (2014b) that the resulting algorithm can be trapped in local optima, especially in high-dimensions. This is likely due to the impressive cardinality of $\mathcal{Z}$. An acceptance-rejection algorithm, performing stochastic optimization along the lines of a general proposal by Chakraborty and Chaudhury (2008), is described in Farcomeni (2014b). A similar algorithm was used in a completely different context by Farcomeni and Viviani (2011).

The algorithm proceeds by iteratively updating an initial solution, identified by a $Z(0) \in \mathcal{Z}$. Let $Z(t)$ denote the update obtained at the tth iteration. We can define $l(Z(t)) = \sum_{i=1}^{n} \min_{1 \leq c \leq k} \sum_{j=1}^{d} z_{ij}(t)(x_{ij} - \mu_{cj})^2$ to be the loss associated with $Z(t)$. It is easily shown, and actually also implied by the snipping rationale, that in order to minimize $l(Z(t))$ in μ_c, $c = 1, \ldots, k$ one shall obtain μ_c as the centroid of observations within the cth cluster, but using only the dimensions corresponding to a positive entry in $Z(t)$.

Each iteration of our proposed algorithm proceeds by building a candidate new solution Z. The candidate is rejected or accepted with a certain probability. If it is accepted, $Z(t)$ is updated to Z, otherwise $Z(t) \equiv Z(t-1)$. Formally, Z is built by switching the positions of a zero and a positive entry of $Z(t)$. These two entries are selected uniformly at random. The acceptance probability, derived from the general proposal of Chakraborty and Chaudhury (2008), can be defined as

$$p_t = \min\left(e^{-\frac{\log(t+1)}{const}(l(Z)-l(Z(t)))}, 1\right), \tag{8.7}$$

for some tuning constant *const*. This guarantees that a candidate solution leading to smaller loss is always accepted, while a candidate solution leading to slightly larger loss may also be accepted but with probability being lower and lower as t grows. This allows the algorithm to escape local optima and explore the solution space during the first few runs. The tuning constant *const* is set approximately equal to the maximal expected change in loss when switching two non-contaminated entries. It is recommended in Farcomeni (2014b) that the algorithm is repeatedly run with different values of *const* in order to better tune this parameter. A summary is given in Algorithm 8.4.

In Farcomeni (2014b) it is shown that the proposed algorithm converges to the global optimum as t grows. The rate of convergence is slightly smaller than $1/2$. Reaching the global optimum is guaranteed only when t grows indefinitely; while the number of iterations T is usually fixed in advance, and of course always finite. In Farcomeni (2014b) $T = 2 * 10^5$ is used. A consequence and clear limitation is that we not only know if we have found the global optimum, but we do not even know if we have reached a mode of (8.6) of any kind. An accelerated convergence strategy is described in Farcomeni (2014b), which has the same limitations anyway.

In order to study the robustness properties of snipped k-means, Farcomeni (2014b) adopts the same restrictions of Ruwet *et al.* (2013). It is shown in Farcomeni (2014b) that for snipped k-means both the individual and cell breakdown points, when restricted to well clustered data sets, are bounded from below by ε. Formally, $\varepsilon^{(i)} \geq \varepsilon$ and $\varepsilon^{(c)} \geq \varepsilon$.

Trimmed k-means can achieve then a cell breakdown point equivalent to snipped k-means only when the trimming level is set equal to $p\varepsilon$. This may be too large to be feasible, and is a consequence of the fact that trimming is not tailored for taking care of outliers in isolated entries.

Algorithm 8.4 The snipped k-means algorithm

Update the row memberships
for $i = 1, \ldots, n$ do

Let $d_{ic}^2 = \sum\limits_{j=1}^{p} (x_{ij} - \bar{x}_{cj}(Z(t)))^2$, $c = 1, \ldots, k$.

Let $r_i = \operatorname{argmin}_c d_{ic}^2$.

Set $u_i = 0$. Set $u_{ir_i} = 1$.

end for

Estimate the centroid matrix
For $c = 1, \ldots, k$ estimate $\bar{X}(Z(t))$ as

$$\bar{x}_{cj}(Z(t)) = \sum_{i=1}^{n} \frac{\sum z_{ij}(t) u_{ic} x_{ij}}{\sum_{i=1}^{n} z_{ij}(t)}$$

Current loss
$l(Z(t)) = \sum_{i=1}^{n} \min\limits_{1 \le c \le k} \sum_{j=1}^{p} z_{ij}(t)(x_{ij} - \bar{x}_{cj}(Z(t)))^2$

Candidate solution
Switch a zero and a one of $Z(t)$ uniformly at random to obtain Z
For $c = 1, \ldots, k$ estimate $\bar{X}(Z)$ as

$$\bar{x}_{cj}(Z) = \sum_{i=1}^{n} \frac{\sum z_{ij} u_{ic} x_{ij}}{\sum_{i=1}^{n} z_{ij}}$$

$l(Z) = \sum_{i=1}^{n} \min\limits_{1 \le c \le k} \sum_{j=1}^{p} z_{ij}(x_{ij} - \bar{x}_{cj}(Z))^2$

Acceptance/rejection
Accept Z with probability p_t as in (8.7).
if Z is accepted then

$Z(t+1) = Z$, $\bar{X} = \bar{X}(Z)$, $U = U(Z)$.

else

$Z(t+1) = Z(t)$

end if

8.4 Choosing the trimming and snipping levels

Now that we have reviewed many robust data reduction methods that are based on trimming or snipping, a natural open issue is the practical choice of the trimming or snipping level ε. We have worked treating ε as known, or at least fixed in advance, so far.

We give some ideas in this section, which apply both to distance based and to model based approaches which will be discussed in the next chapter.

In order to give an idea of the issues faced when selecting ε, we point out that the objective function of distance based approaches and maximum of the likelihood in model based ones are non-decreasing in ε. Given that the number of parameters is fixed, a greedy approach would obviously lead to fix $\varepsilon = 100\%$, that is, not to perform sample reduction at all. This is of course meaningless, and only tells us that ε cannot be fixed by optimizing the same criterion used for estimation of the other parameters.

One obvious possibility is to *penalize* the objective function, for instance adding a term which is a non-decreasing function of ε. This idea has not been explicitly pursued in the literature so far, and would lead to a unified optimization problem in which all parameters, including ε, are simultaneously estimated. A clear limitation is that we would trade the problem of tuning ε with the problem of choosing a penalty (or tuning parameters involved in a special form of a penalty function). Another possibility is to either visually or formally evaluate the optimum of the objective function and/or the stability of the results for a grid of snipping/trimming levels ε. This less formal approach has been pursued in the literature in different ways (e.g, Farcomeni (2009b), García-Escudero *et al.* (2011)), and we briefly describe it below.

In our experience, an extremely careful choice of ε is not crucial. Given its definition, it shall be set approximately equal to the expected contamination level. If the trimming or snipping level is slightly larger than the contamination actually present in the data, all contamination can be excluded. If ε is slightly too large there will be a loss of efficiency. In the context of cluster analysis ($k > 1$), this is slightly less important that the case of location and scatter estimation discussed in Chapter 2, corresponding to $k = 1$. Some clean entries (or observations) will be discarded, but this should not affect substantially the parameter estimates. With snipping procedures, discarding some clean entries should not even affect the quality of clustering, as snipped observations will still be assigned to a cluster, and most likely the same cluster they would have been assigned to without snipping clean entries. Even if ε is too large, it is very unlikely that an entirely clean observation is substantially snipped and for moderate p trimming is almost impossible. When using trimming procedures and ε is too large, some clean observations will be discarded together with the contaminated ones, but otherwise there will not be other deleterious effects apart from a slight loss of efficiency.

Note that there practically is no need of adjustment of centroid estimates because of trimming. As far as dispersion parameters are concerned, on the other hand, there may be issues of consistency and a consistency factor is often used in the literature when $k = 1$. This was established in Tallis (1963) and Butler *et al.* (1993), see also for instance Pison *et al.* (2002) and Cerioli *et al.* (2014) for recent applications. Given that our focus is mostly on centroid estimation and only comparative assessment of dispersion among groups, we do not pursue this issue anymore, and discuss it slightly more in depth in Chapter 11. Note furthermore that to the best of our knowledge this problem has not been yet investigated for the relevant case in which $k > 1$. We speculate that consistency factors might be slightly different after incorporating label uncertainty brought about by unsupervised learning.

The consequences of setting ε slightly too small are much more important, as some contaminated entries will not be discarded and will affect the estimates. Nevertheless, gross outliers which are mostly influential will always be the first to be discarded, so that if a small fraction of outliers is not flagged, these will likely be those corresponding to less influential contamination.

We begin by reviewing the G statistic of Farcomeni (2009b), adapting it to our context. The G statistic is an empirical evaluation of the breakdown point, as it is a function of the difference between the parameter estimates with or without flagging outliers. Formally,

$$G(\varepsilon, 0) = \max ||(\theta(\varepsilon) - \theta(0))|| / \max(||\theta(\varepsilon)||, ||\theta(0)||), \qquad (8.8)$$

where all operators are entry-wise and θ is a short-hand notation for the parameters involved in the model (excluding V and Z). We denote with $\theta(\varepsilon)$ the estimates obtained with trimming or snipping level ε, and with $\theta(0)$ the classical estimates. Note that some form of matching must precede (8.8), as label switching may have the two parameter estimates correspond to a different equivalent permutation of the cluster labels. In order to overcome this difficulty, Farcomeni (2009b) propose to match centroid estimates by their relative magnitude, that is, to sort centroid estimates. After matching, the relative difference in parameter estimates with a trimming or snipping level ε is evaluated, and the G statistic is set to the maximal relative difference observed vector-wise.

The idea of Farcomeni (2009b) can be extended in full generality by expressing the statistic as the difference of any function of the estimates when comparing two snipping/trimming levels.

A generalized G statistic can be simply defined as expressing the difference between estimates for different ε:

$$G(\varepsilon_1, \varepsilon_2) = f(\theta(\varepsilon_1), \theta(\varepsilon_2)), \qquad (8.9)$$

where f is some function, and $\theta(\cdot)$ this time refers to the entire set of estimated parameters (including therefore V and Z). Farcomeni (2009b) proposed using the maximum relative difference in (8.9), but any distance function between the two estimates would be effective. The expression (8.8) is obtained when $\varepsilon_2 = 0$. Furthermore, also any statistics evaluating the quality of clustering, such as the average Silhouette, would be effective.

Another possibility mentioned in Farcomeni (2009b) is to measure the agreement between the labels obtained based on ε_1 and ε_2. One can use Cohen's kappa as in Reilly *et al.* (2005), Rand's C_k as in Chae *et al.* (2006), or the modified Rand index (Hubert and Arabie, 1985). A final idea is to use the objective function itself, which can be scaled if needed. This idea applies particularly well to model based clustering, where the objective function is the likelihood function; making (8.9) a function of popular goodness of fit measures based on the log-likelihood.

Algorithm 8.5 Choosing ε

Choose a stopping value δ
Choose a grid of values $\varepsilon_1 < \ldots < \varepsilon_T$, with $\varepsilon_0 = 0$.
Initialize $G = 0$, $j = 1$
while $G(\varepsilon_j, 0) - G > \delta$ **and** $j <= T$ **do**
   Set $G = G(\varepsilon_j, 0)$.
   $j = j + 1$
end while
Set $\varepsilon = \varepsilon_{j-1}$

    The G statistic can be used to evaluate whether there is any contamination at all. If there is no contamination, $G(\varepsilon, 0)$ will be small regardless of ε. The G statistic can be plot for a grid of values, and ε can be increased until there is no additional substantial advantage in an increase. A formal procedure along these lines is given in Algorithm 8.5.

    The principle behind the proposed strategy is that if when increasing ε_j to ε_{j+1} additional outliers are identified, there will be a relatively large change in at least one of the entries of θ, say bigger than δ. Otherwise, there will be a small (lower than δ) change and there is no need to increase ε. The choice of δ depends on the application and on the sample size. We suggest choosing smaller values of δ for large matrices, since a single outlier may not yield a big change in the centroid matrix in the presence of many observations. Given that overestimation of ε is less problematic than underestimation, it may be wiser is some cases to proceed in a backward rather than in a forward fashion.

    The trimmed likelihood curves of García-Escudero *et al.* (2011) can be used instead to choose k. These curves are based on plotting the likelihood obtained at convergence against ε, for different choices of k. The number of groups shall be set equal to the minimum k for which there is no substantial increase in the maximum of the likelihood when adding one extra group. Note that the trimmed likelihood curves can also be used for simultaneous selection of k and ε, if desired. We will illustrate trimmed likelihood curves in some of the examples. Generalization of trimmed likelihood curves to the case of distance based clustering is straightforward, as one can simply substitute the likelihood with any general objective function.

8.5   Examples

We provide in this section examples of robust model based clustering techniques.

TABLE 8.2

Metallic oxide data. Comparison of the modified Rand index ($\cap$ Rand is based on samples that are not entirely discarded by any of the procedures, Rand on samples that are not discarded only by the procedure itself) and number of trimmed (trim) samples. A comparison across different trimming levels is not reliable since the Rand index is computed on a different number of elements.

ε	tkmeans(α)			skmeans(ε)		
	$\cap$ Rand	Rand	trim	$\cap$ Rand	Rand	trim
0.05	0.17	0.17	2	0.17	0.19	1
0.075	0.22	0.22	3	0.22	0.19	1
0.10	0.20	0.20	4	0.20	0.22	3
0.15	0.18	0.18	5	0.18	0.22	3
0.175	0.15	0.15	6	0.15	0.18	5

8.5.1 Metallic oxide data

We begin working with the metallic oxide data. We have already seen that classical clustering methods give badly behaved solutions, with very small Rand index. We now apply robust clustering methods, which in principle should not be affected by contamination. We assume the number of clusters is known, and fix $k = 2$.

We begin using PAM, which gives exactly the same solution as 2-means, with lots 6 and 7 of Type 2 belonging to one group and all the other rows belonging to the other group. The resulting Rand index is a disappointing 5.8%.

We now proceed by choosing the trimming and snipping levels. We use the G statistic to this end. We set f as in (8.9) so to use only the mean square difference between centroid estimates. This choice and $\delta = 10\%$ leads to select $\varepsilon = 10\%$ both for trimmed and snipped k-means.

We now use different trimming and snipping levels, whose results are reported in Table 8.2. Note that this part of the analysis is used only as a global appraisal of the performance of robust clustering methods and of the contamination. Our formal analysis will follow. In Table 8.2, the first column gives ε used both for the trimmed and snipped k-means. We compare the true and predicted group labels using the modified Rand index (Hubert and Arabie, 1985). The column with $\cap$ gives modified Rand index computed using samples that are not entirely discarded by any of the procedures. The "Rand" column gives instead the modified Rand index computed on samples that are not discarded only by the procedure itself. We also report the number of samples entirely discarded (*trim*).

From Table 8.2 we can see that the Rand index of the entries that are not flagged by any of the procedures is always the same. On the other hand, the actual Rand index of snipped k-means is slightly larger than that of trimmed

TABLE 8.3
Handwritten digits data. Modified Rand index and conditional probability of correct classification $\Pr(j|j)$, $j = 0, \ldots, 9$, for different clustering procedures.

Method	Rand	0	1	2	3	4	5
k-means	67%	0.37	0.76	0.88	0.71	0.43	0.97
PAM	64%	0.57	0.89	0.89	0.78	0.72	0.98
tkmeans(5%)	71%	0.64	0.90	0.88	0.89	0.80	0.99
skmeans(5%)	83%	0.87	0.91	0.91	0.95	0.76	0.99
tkmeans(10%)	75%	0.61	0.93	0.92	0.92	0.83	0.99
skmeans(10%)	85%	0.93	0.93	0.92	0.95	0.80	0.99

Method	Rand	6	7	8	9
k-means	67%	0.95	0.82	0.44	1.00
PAM	64%	0.95	0.44	0.75	0.99
tkmeans(5%)	71%	0.97	0.86	0.67	0.99
skmeans(5%)	83%	0.99	0.90	0.85	0.99
tkmeans(10%)	75%	0.98	0.84	0.79	1.00
skmeans(10%)	85%	0.99	0.90	0.82	0.99

k-means in all cases except $\varepsilon = 7.5\%$. Furthermore, snipped k-means is able to retain and cluster one or two additional samples in all cases. For the chosen level of $\varepsilon = 10\%$ and snipped k-means we have three structural outliers and two component-wise outliers.

8.5.2 Handwritten digits data

We now work with the handwritten digits data.

In Table 8.3 we report the modified Rand index and, for each digit, the conditional probability of correct classification for different procedures. The latter is obtained dividing the diagonal of the classification matrix by the frequency table of true labels. The operation is intended entry-wise. Cluster labels are permuted in order to maximize the sum of the elements in the diagonal of the classification matrix. We use $\varepsilon = 5, 10\%$.

From Table 8.3 we can see that with the same number of discarded entries snipping can achieve a Rand index which is substantially larger than trimmed k-means, due to the additional flexibility in arranging them. There is a substantial improvement also for what concerns the misclassification error. We remark that the Rand index is based on different sets of observations for different ε, and hence should not be compared across different values of this tuning parameter.

The difference in misclassification error among the procedures considered is substantial. There is strong evidence of presence of structural outliers and of component-wise outliers. The most important example is given by the 0, which seems to be much better classified with snipping. This is probably because minimal scratches of the handwriting can make it look like a 6, a 9 or even an 8. Hence there are many component-wise outliers in connection with the digit 0. Trimming on the other hand seems to be particularly useful for the digit 4, which is the only correctly classified 3-4% less often with snipping than with trimming. This probably happens since by removing some entries few handwritten 4 can look like a 1. This is a very tricky situation in which structural outliers may be misclassified by snipped k-means because by removing few entries, the remaining ones could be assigned to the wrong group.

8.1 R Illustration (Robust clustering metallic oxide data) *For the metallic oxide example, data are stored in a data frame* `fell`. *PAM was obtained simply with a call to function* **pam**. *The G statistics (and Rand index) for trimmed and snipped k-means at a grid of values* `eps` *were obtained with the following code:*

```
> library(tclust)
> source("rand.index.r")
> library(snipEM)
>
> eps=c(0.05,0.075,0.1,0.15,0.175,0.2,0.25)
>
> res=matrix(NA,length(eps),7)
> res[,1]=eps
>
> gstat=matrix(NA,length(eps),2)
> g0=kmeans(fell[,-1],2)$cent
> g0=g0[order(g0[,1]),]
>
> i=1
> for(ep in eps) {
+ tk=tkmeans(fell[,-1],2,alpha=ep)
+ V=matrix(1,31,8)
+ V[tk$clust==0,]=0
+ R=tk$clust
+ R[R==0]=sample(2,sum(R==0),replace=TRUE)
+ sk=skmeans(as.matrix(fell[,-1]),2,V,R)
+
+ gstat[i,1]=mean((t(tk$cent[,order(tk$cent[1,])])
+ -g0)^2)
+ gstat[i,2]=mean((sk$mu[order(sk$mu[,1]),]-g0)^2)
+
+ u1=tk$clust!=0
+ u2=apply(sk$V,1,sum)>0
```

```
+ u=u1 & u2
+
+ res[i,-1]=c(rand.index(unclass(fell[u,1])),
+ tk$clust[u]),
+ rand.index(unclass(fell[u1,1]),tk$clust[u1]),
+ sum(tk$clust==0),rand.index(unclass(fell[u,1]),
+ sk$clust[u]),rand.index(unclass(fell[u2,1]),
+ sk$clust[u2]),sum(apply(sk$V,1,sum)==0))
+ i=i+1}
```

9

Robust Model-Based Clustering

The focus of this chapter is on model based clustering (Fraley and Raftery, 2002) through finite mixture models. We shall begin by clarifying that there are now many approaches to robust finite mixture modeling. Some of them are based on using flexible components in the mixture, like Banfield and Raftery (1993), McLachlan and Peel (2000), Frühwirth-Schnatter and Pyne (2010). It is common to replace the usual assumption that each mixture component follows a Gaussian distribution with one according to which one or more (or even all) follow more flexible distributions. These may have heavier tails than the Gaussian, in order to make extremes less unusual. A common choice is the multivariate T distribution. Skewed distributions instead can be used to accommodate unusual cluster shapes, and so on. Another possibility is to use one or more additional components to accommodate an outlier generating distribution of some kind. Contamination by background noise, for instance, is often formally defined using multivariate uniform distributions (often, independent in each component). A related approach is that of Fraley and Raftery (1998), who propose a Gaussian mixture with an additional component modeled as a Poisson process to handle noisy data. Flexible mixtures are very interesting but in many cases lack formal robustness properties, that is, it can be easily shown that the global and local robustness properties are very similar to those of the conventional Gaussian mixture models. An underlying difference with other robust methods is that contamination is often deemed in robust statistics as being unusual under the assumed model. Hence an "anticipated" contamination is not contamination, but simply a non-Gaussian model expected for some of the observations.

There are also other approaches to robust mixture modeling, based on robustifying the estimation procedure, for instance, by downweighting the unusual observations. Examples can be found for instance in Cambpell (1984) and Markatou (2000). In this chapter we will review only approaches giving a *zero* weight to unusual observations. These are formally defined as trimming (Gallegos and Ritter (2005), Neykov *et al.* (2007), García-Escudero *et al.* (2008), Gallegos and Ritter (2009a, 2010)) when the entire outlier is zero weighted; or snipping (Farcomeni, 2014a), when possibly only some entries are discarded.

As in the previous chapter, we deal with component-wise and structural contamination separately. It is important here to mention that trimming and snipping methods can be directly combined in a unified algorithm when both are present. As a matter of fact, one can prescribe a proportion of trimmed observations ε_1 and a proportion of snipped entries ε_2 and quite easily take care of the simultaneous presence of structural and cell-wise outliers. This has not been explored in the robust clustering literature so far. A snipping and trimming estimator of location and scatter, that is, for the case $k = 1$ along these lines is available from package snipEM as function stEM, and briefly mentioned in Farcomeni (2014a) and Farcomeni (2015).

One possibility of interpreting some robust distance based methods is that they correspond to model based clustering methods with the additional assumption that the covariance matrices are homogeneous across groups, and diagonal. Most of the material in this chapter can be seen as a generalization of that of the previous chapter. We have chosen to repeat most of the reasoning from scratch to make this chapter self contained and to help the reader better understand the material already discussed.

9.1 Robust heterogeneous clustering based on trimming

We assume that $n\varepsilon$ observations are contaminated, for some (fixed and known) $\varepsilon > 0$. If we let z_i denote an indicator of clean observations, the constraint that there are exactly $n\varepsilon$ outliers can be expressed as $\sum_i (1 - z_i)/n = \varepsilon$. If the ith observation is contaminated, it is generated from an almost arbitrary subject-specific distribution $g_i(\cdot)$. Some assumptions, which are slightly general, are needed in order to make the model identifiable and maximization of the likelihood possible. These are briefly outlined below.

The data generating distribution can be summarized as:

$$f(X_i|c) \sim z_i \phi_p(X_i, \mu_c, \Sigma_c) + (1 - z_i)g_i(X_i), \tag{9.1}$$

that is, we have k clusters and $\sum(1 - z_i)$ observations arbitrarily placed in $\mathbb{R}^p$. That is, each cluster is made of a majority of observations generated from a Gaussian distribution with cluster-specific mean μ_c and covariance matrix Σ_c, $c = 1, \ldots, k$; and a small fraction of structural outliers.

We also assume a prior probability π_c, $c = 1, \ldots, k$, that X_i belongs to the cth cluster, with obviously $\pi_c > 0$ and $\sum_{c=1}^{k} \pi_c = 1$. Assumptions on the moments are that $||\mu_c - \mu_d|| > 0$ when $c \neq d$; and that the covariance matrices Σ_c are strictly positive definite. A final assumption regards the contaminating distributions $g_i(\cdot)$. We assume that contaminated entries are generated far enough from the clean model and are affine independent. These conditions were studied in detail in Gallegos and Ritter (2005) and García-Escudero *et al.* (2008). These assumptions are seldom restrictive, and mostly formalize the concept of outliers. As a matter of fact, if an observation is not separated from the Gaussian component it can be perceived as being generated from the clean data distribution, and hence is not an outlier. The resulting model is very similar to the spurious outliers model of Gallegos and Ritter (2005). Additional conditions usually deal with how groups have to be separated in order to have a well-behaved cluster structure. These assumptions are often needed in order to derive results on the breakdown properties of the procedures.

The likelihood corresponding to model above can be computed as

$$\prod_c \prod_i (\pi_c \phi_p(x_i, \mu_c, \Sigma_c))^{z_i u_{ic}} \prod_i g_i(x_i)^{1-z_i}. \tag{9.2}$$

There are two main differences between (9.1) and flexible mixture models. The first is that $g_i(\cdot)$ is left almost completely unspecified. The second is that under the separation condition, the likelihood (9.2) is maximized by using point masses on the contaminated observations, that is, a profile likelihood is obtained by estimating $\hat{g}_i(x_i) = 1$. This is related to non-parametric maximum likelihood estimation, see also Farcomeni (2014a) on this point. This fact is presented in García-Escudero *et al.* (2008) as the possibility to ignore the nonregular entries in (9.2), and the separation condition is formally expressed as equivalent to the possibility of ignoring the contribution of contaminated observations, that is,

$$\arg\max_z \sup_{\mu, \Sigma, \pi} \prod_c \prod_i (\pi_c \phi_p(x_i, \mu_c, \Sigma_c))^{z_i u_{ic}} \subseteq \arg\max_z \prod_i (g_i(x_i))^{1-z_i}.$$

The expression above indicates that the maximum of the likelihood with respect to z is smaller than or equal to the maximum obtained by separately maximizing the contributions of the clean and contaminated observations. The contaminating observations therefore disappear from (9.2) and we maximize the resulting profile likelihood, which is based only on the clean data:

$$\prod_c \prod_i (\pi_c \phi_p(x_i, \mu_c, \Sigma_c))^{z_i u_{ic}}. \tag{9.3}$$

The latter shall be maximized in z, U, π, μ, and Σ, with constraints on Σ as outlined below.

9.1.1 A robust CEM for model estimation: the `tclust` algorithm

Maximization of (9.3) is set up through a classification expectation maximization (CEM) type algorithm (Celeux and Govaert, 1992). The classical unconstrained CEM approach is seen to often lead to spurious solutions in which one or more estimates $\hat{\Sigma}_c$, $c = 1, \ldots, k$ break down. A nice discussion about this problem, with some examples to clarify the ideas, can be found in García-Escudero *et al.* (2008, 2014). See also Maronna and Jacovkis (1974).

The use of a vector of weights π_c, $c = 1, \ldots, k$ in connection with the classification likelihood leads to slightly more unbalanced cluster cardinalities with respect to the classical assumption that $\pi_c = 1/k$. This is also known as penalized CEM, see for instance Symons (1981) and Bryant (1991).

In order to avoid spurious solutions, estimates of Σ_c, $c = 1, \ldots, k$ shall be constrained (see, e.g., Hathaway (1985)). There are many possible constraints available in the robust clustering literature. One possibility is to constrain the cardinality of the clusters as in Gallegos and Ritter (2010). A more direct but less simple constraint is to place bounds on the determinant of the covariance matrices, as in Gallegos and Ritter (2009a,b). Constraints on the determinants are rather natural and have clear advantages, but are slightly difficult to impose within the CEM algorithm. A third class of constraints is related to bounding the *ratios* of the eigenvalues of the covariance matrices, so that they are somehow pooled together and constrained to be similar at least up to a point. This eliminates extreme situations in a natural way, with basically the only disadvantage that affine equivariance is lost. Formally, the approach as proposed by García-Escudero *et al.* (2008, 2014) is to fix a constant $const \geq 1$ and constrain

$$\frac{\max_j \lambda_1(\Sigma_j)}{\min_j \lambda_p(\Sigma_j)} \leq const, \tag{9.4}$$

where $\lambda_1(A)$ and $\lambda_p(A)$ denote the largest and smallest eigenvalue of A. Affine equivariance is lost as the constraint is not invariant to scale transformations. A constraint of the kind (9.4) has got two advantages in our opinion: first of all, it can easily be imposed (Fritz *et al.*, 2013) within the estimation algorithm. Secondly, it can be easily tuned. The extreme case $const = 1$ corresponds to the case in which all covariance matrices are equal. We mention here that affine equivariance may not even be so important in clustering applications, as outlined in Hennig and Liao (2013).

We can now set up a CEM algorithm for trimmed robust clustering, which is usually referred as the `tclust` algorithm (Fritz *et al.*, 2012).

At the CE step we maximize (9.3) with respect to U. First of all we compute for $c = 1, \ldots, k$

$$D_i(c) = \log(\pi_c \phi_p(x_i, \mu_c, \Sigma_c)). \tag{9.5}$$

We then set $u_{i\tilde{c}_i} = 1$ where $\tilde{c}_i = \arg\max_c D_i(c)$ for all i, and zero otherwise. The basis behind (9.5) is that each subject is assigned, for fixed μ and Σ, to the latent class which leads to maximization of the likelihood. The exponential of (9.5), opportunely rescaled, can also be seen as an estimate of the posterior probability that the ith subject belongs to the cth cluster.

The main difference between the `tclust` algorithm and classical CEM algorithms for mixture models is that there is the inclusion of a concentration step in which some observations are trimmed. This is performed by setting $z_i = 0$ for the $n\varepsilon$ smallest values of $D_i(\tilde{c}_i)$, $i = 1, \ldots, n$; and $z_i = 1$ otherwise. Note that this is absolutely equivalent to setting $u_i = 0$ for all observations for which $z_i = 0$.

We then proceed with an M-step, in which (9.3) is optimized with respect to μ, Σ and π conditionally on the current values for z and U. Closed form solutions are available for π and μ: we have that $\pi_c \propto \sum_i u_{ic}$, while for $c = 1, \ldots, k$ and $j = 1, \ldots, p$ we have that

$$\mu_{cj} = \frac{\sum_i y_{ij} z_i u_{ic}}{\sum_i z_i u_{ic}}. \tag{9.6}$$

An initial estimate of Σ_c is obtained by fixing

$$\Sigma_c = \frac{\sum_i z_i u_{ic}(y_i - \mu_c)(y_i - \mu_c)^T}{\sum_i z_i u_{ic}}. \tag{9.7}$$

This initial estimate is then modified in order to satisfy the constraint (9.4), if needed. Fritz *et al.* (2013) solve (9.7) after restriction to truncated eigenvalues. The covariance matrices obtained with (9.7) are expressed as $V_c' \Lambda_c V_c$, $c = 1, \ldots, k$. The eigenvalues are then truncated using an unknown bound θ, so that λ_c is replaced with θ whenever $\lambda_c < \theta$; and λ_c is replaced with *const*θ whenever $\lambda_c > const\theta$. It is straightforward to check that this truncation recovers a feasible solution of (9.4) for any θ. Approximate optimization with respect to Σ_c under (9.4) is therefore achieved by maximization with respect to θ. It is shown in Fritz *et al.* (2013) that this is equivalent to minimize in θ the expression

$$\sum_{c=1}^{k} \sum_{i=1}^{n} u_{ic} \sum_{j=1}^{p} \left(\log(\lambda_{jc}(\theta)) + \frac{\lambda_{jc}}{\lambda_{jc}(\theta)} \right),$$

where λ_{jc} is the jth eigenvalue of (9.7), and $\lambda_{jc}(\theta)$ its truncated version. To this end, it is shown in Fritz *et al.* (2013) that $2kp + 1$ function evaluations suffice to impose the eigenvalue constraint, and optimizing at least approximately the objective function involved. A more complex algorithm is proposed in García-Escudero *et al.* (2008), adapting methods from Dykstra (1983).

We summarize the `tclust` algorithm as Algorithm 9.1 for the CE and concentration steps, and Algorithm 9.2 for the M step.

9.1.2 Properties

It can be formally argued that the `tclust` method, that is, robust model based clustering with impartial trimming, has good robustness behavior and nice theoretical properties.

The latter are outlined in García-Escudero *et al.* (2008), where the existence of the MLE is established for any $n > 0$, together with the existence of the MLE for the population version of the likelihood. More importantly, it is shown that the MLE is consistent to the population solution.

Robustness properties are studied in García-Escudero *et al.* (2008), Ruwet *et al.* (2012), Ruwet *et al.* (2013), and Farcomeni (2014a). In Ruwet *et al.* (2012), the influence functions of the `tclust` procedure are evaluated only for the case case of $p = 1$ and $k = 2$. The influence functions are derived

Algorithm 9.1 Robust classification EM algorithm (`tclust`): CE and concentration steps

---

Classification E-step
for $i = 1, \ldots, n$ **do**

$$\tilde{c}_i = \arg\max_c \pi_c \phi_p(x_i, \mu_c, \Sigma_c),$$

$u_{i\tilde{c}_i} = 1$
for $c \neq \tilde{c}_i$ **do**
$\quad u_{ic} = 0$
end for
end for
Concentration step
Let

$$\pi_{\tilde{c}_{i_1}} \phi_p(x_i, \mu_{\tilde{c}_{i_1}}, \Sigma_{\tilde{c}_{i_1}}) \leq \pi_{\tilde{c}_{i_2}} \phi_p(x_i, \mu_{\tilde{c}_{i_2}}, \Sigma_{\tilde{c}_{i_2}}) \leq$$
$$\leq \cdots \leq \pi_{\tilde{c}_{i_n}} \phi_p(x_i, \mu_{\tilde{c}_{i_n}}, \Sigma_{\tilde{c}_{i_n}})$$

for $j = 1, \ldots, n\varepsilon$ **do**
$\quad z_{i_j} = 0$
end for
for $j = n\varepsilon + 1, \ldots, n$ **do**
$\quad z_{i_j} = 1$
end for

---

explicitly and it is shown that they are bounded in many cases. Surprisingly, the influence functions of the cluster centers are not linear. All are continuous, except at cluster boundaries.

As far as global robustness is concerned, first of all it is clear that estimates for π can not break down as they are bounded. Appropriate definitions of breakdown points for bounded parameters can be found in Genton and Lucas (2003), but a study of breakdown of π was never pursued in the literature so far. Universal individual breakdown points for Σ_c are derived in Ruwet *et al.* (2013), where it is shown that if $\varepsilon \leq .5 - k(p+1)/2n$, $\varepsilon^{(i)} \geq \varepsilon$.

As far as centroid estimates are concerned, the universal breakdown points are infinitesimal. In order to obtain useful results, Ruwet *et al.* (2013) restrict to a class of well-clustered data sets. This is defined in technical detail in Ruwet *et al.* (2013), and involves lower bounds on the number of observations arising from each cluster, and on distances among cluster centers. If clean data arises within that class, and

$$\varepsilon \leq \min\left(.5 - \frac{k(k-1)p}{2n} + \frac{1}{2n}, \frac{1}{k} - \frac{1}{n}\right),$$

the restricted individual breakdown point for μ_c is such that $\varepsilon^{(i)} \geq \varepsilon$. A

Algorithm 9.2 Robust classification EM algorithm (`tclust`): M-step

M-step
  for $c = 1, \ldots, k$ **do**

$$\mu_c = \frac{\sum_i z_i u_{ic} x_i}{\sum_i z_i u_{ic}},$$

$$\Sigma_c = \frac{\sum_i z_i u_{ic}(x_i - \mu_c)(x_i - \mu_c)^T}{\sum_i z_i u_{ic}},$$

  Obtain

$$\Sigma_c = V_c^T \Lambda_c V_c.$$

  end for
  if $\frac{\max_{jc} \lambda_{jc}}{\min_{jc} \lambda_{jc}} < const$ **then**
    Set $\hat{\theta}$ as the minimizer of

$$\sum_{c=1}^{k} \sum_{i=1}^{n} u_{ic} \sum_{j=1}^{d} \left(\log(\lambda_{jc}(\theta)) + \frac{\lambda_{jc}}{\lambda_{jc}(\theta)} \right)$$

  for $c = 1, \ldots, k$ **do**

$$\Sigma_c = V_c^T \Lambda_c(\theta) V_c,$$

    where $\Lambda_c = \operatorname{diag}(\lambda_{jc}^m)$.
  end for
  end if

---

discussion of the cell breakdown points for the `tclust` procedure can be found in Farcomeni (2014a), who obtains quite analogous results. It can be shown that the restricted cell breakdown point for $\hat{\mu}_c$ is such that $\varepsilon^{(c)} \geq \frac{\varepsilon}{p}$; while for the scatter estimator we also have the universal cell breakdown is such that $\varepsilon^{(c)} \geq \frac{\varepsilon}{p}$.

9.2 Robust heterogeneous clustering based on snipping

Robust model based clustering can also be performed under an entry-wise spurious outliers model.

    In this section we put forward a model that is very much along the lines of that in Section 8.3. We in fact assume that $np\varepsilon$ entries of the data matrix X are contaminated, for some $\varepsilon > 0$, and let z_i denote a binary vector such that $\sum_{ij} z_{ij}$ is equal to the integer closest to $np(1 - \varepsilon)$. The indicator z_{ij} identifies whether the $(ij)^{th}$ entry of the data matrix is free of contamination. When $z_{ij} = 0$, the jth element of the ith observation is flagged as being contaminated. We assume a model that is a generalization of (8.5), that is:

$$f(X_i|c) \sim z_i \phi_p(X_i, \mu_c, \Sigma_c) + (1 - z_i)g_i(X_i). \tag{9.8}$$

Model (9.8) is different from (8.5) in that we now have a cluster-specific co-variance matrix Σ_c, $c = 1, \ldots, k$.

We also assume a prior probability π_c, $c = 1, \ldots, k$, that X_i belongs to the cth cluster, with obviously $\pi_c > 0$ and $\sum_{c=1}^{k} \pi_c = 1$. Note that this prior probability is defined *conditionally* on X_i not being entirely contaminated. Assumptions on the moments are that, clearly, $||\mu_c - \mu_d|| > 0$ when $c \neq d$; and that the covariance matrices Σ_c are strictly positive definite. The spurious outliers model of Gallegos and Ritter (2005), and discussed in the previous section, is obtained with the additional constraint that $\sum_j (1 - z_{ij}) \sum_j z_{ij} = 0$ for all $i = 1, \ldots, n$. This constraint simply implies that if an entry is contaminated, then also all other entries in the same row must be. Model (9.8) generalizes therefore model (9.1) to the case of entry-wise contamination (where also structural outliers may be present); and generalizes model (8.5) to the case of cluster specific covariance matrices.

The likelihood corresponding to model (9.8) can be computed as

$$\prod_c \prod_i (\pi_c \phi_{\sum_j z_{ij}}(x_{i(z_i)}, \mu_{c(z_i)}, \Sigma_{c(z_i)})g_i(x_{i(1-z_i)}))^{u_{ic}} \prod_i g_i(x_i)^{1 - \sum_c u_{ic}}. \tag{9.9}$$

In (9.9), multivariate densities of varying dimensionality are involved. We use the notation $\phi_{\sum_j z_{ij}}$ to indicate the density of a multivariate normal of dimensionality $\sum_j z_{ij}$. In the limiting case in which $\sum_j z_{ij} = 1$, ϕ_1 denotes the density of a univariate Gaussian distribution. When $\sum_j z_{ij} = 0$, the ith observation is a structural outlier, arising completely from $g_i(\cdot)$. Further, with $\mu_{c(z_i)}$ we denote the entries of μ_c that correspond to a nonzero element in the vector z_i, the same applies to $\Sigma_{c(z_i)}$, and x_{z_i}. This approach corresponds to a situation in which elements corresponding to a zero entry in z_i are missing at random, and are therefore ignored for estimation of μ and Σ. It is demonstrated in Farcomeni (2014a) that the problem is well defined as long as two conditions are satisfied:

$$\sum_i z_{ij} u_{ic} > 0 \quad \forall j = 1, \ldots, p; c = 1, \ldots, k; \tag{9.10}$$

and

$$\sum_i u_{ij} u_{ih} z_{ic} > 0 \quad \forall j, h = 1, \ldots, p; c = 1, \ldots, k. \tag{9.11}$$

The first condition guarantees that there is at least one clean observation for each dimension and cluster; while the second condition guarantees identifiability of Σ_c, with a similar reasoning.

Finally, as with the `tclust` algorithm, we have separation conditions on $g_i(\cdot)$. In our case, the classical separation condition on $g_i(\cdot)$ becomes

$$\arg\max_Z \sup_{\mu,\Sigma,\pi} \prod_c \prod_i (\pi_c \phi_{\sum_j z_{ij}}(X_{i(z_i)}, \mu_{c(z_i)}, \Sigma_{c(z_i)}))^{u_{ic}}$$
$$\subseteq \arg\max_Z \prod_c \prod_i (g_i(x_{i(1-z_i)}))^{u_{ic}} \prod_i g_i(x_i)^{1-\sum_c u_{ic}}.$$

The separation condition above guarantees that the likelihood can be maximized ignoring the contribution of contaminated entries, that is, that the MLE of (9.9) corresponds to the maximizer of

$$\prod_c \prod_i (\pi_c \phi_{\sum_j z_{ij}}(x_{i(z_i)}, \mu_{c(z_i)}, \Sigma_{c(z_i)}))^{u_{ic}}; \qquad (9.12)$$

which can be seen as a profile likelihood.

It is discussed in Farcomeni (2014a) that additional constraints on Z can be used to obtain slightly different contamination models. One can for instance assume that $\sum_j z_{ij} > 0$ for all i, so that no observation is entirely contaminated. This may be useful in applications where clustering *all* of the available observations is important. Similarly, a fixed number of entries can be assumed to be structural outliers, and entirely trimmed by the algorithm.

Before we proceed with estimation of π, μ, Σ, U and Z; we shall notice that, as for snipped k-means, the component-wise outlier model outlined above is not scale invariant: linear transformations propagate the outliers (Alqallaf *et al.*, 2009). Constraints of the kind (9.4) are used to guarantee that the covariance estimates are well defined, but they are also not scale invariant. Hence, with robust model based clustering, the loss of scale invariance when using snipping instead of trimming is much less important, as also trimmed solutions are not scale invariant in the first place.

9.2.1 A robust CEM for model estimation: the `sclust` algorithm

In order to proceed with cluster analysis the maximizer of (9.12) in terms of Z, U, π, μ and Σ is needed, under constraints (9.10) and (9.11) for identifiability. As with the `tclust` algorithm, constraints on the covariance matrices must also be imposed to avoid spurious solutions.

Farcomeni (2014a) proposes to work with the observed likelihood, with U therefore being regarded as a latent random variable. An EM is consequently set up in Farcomeni (2014a). Here, in parallel with the `tclust` algorithm, we set up a classification expectation maximization (CEM) type algorithm (Celeux and Govaert, 1992). Therefore, as with the `tclust` algorithm, U is a parameter. The CEM incorporates a constrained stochastic optimization step for updating Z and a constrained maximization step for updating Σ. The former is used to obtain the optimal feasible snipping configuration, while the latter to obtain locally optimal parameter estimates avoiding spurious solutions.

The algorithm proceeds with a CE step, which involves updating U, a snipping (S) step, which involves updating Z, and a constrained M step for updating the remaining parameters.

As far as the CE step is concerned, it is straightforward to see that for fixed Z, π, μ and Σ the maximizer of (9.12) is obtained by computing

$$D_i(c) = \log(\pi_c \phi_{\sum_j z_{ij}}(x_{i(z_i)}, \mu_{c(z_i)}, \Sigma_{c(z_i)})) \tag{9.13}$$

and setting $u_{i\tilde{c}} = 1$ where $\tilde{c} = \arg\max_c D_i(c)$ for all i such that $\sum_j z_{ij} > 0$. Once again, each subject is assigned, for fixed Z, μ and Σ, to the most likely latent class.

Regarding the S step, a simple stochastic optimization algorithm is set up along the lines of Chakraborty and Chaudhury (2008); Farcomeni (2014a,b). This is similar to the stochastic optimization algorithm outlined for snipped k-means, with the difference that now it is only a step within the CEM algorithm. A sequence of random proposals $Z_c(t)$, $t = 1, \ldots, T$ is built by switching a 0 and a 1 entry uniformly at random in $Z_c(t-1)$. Obviously, $Z_c(0) = Z$, the current optimal solution. The random proposal $Z_c(t)$ is accepted with probability

$$p_t = \min\left(e^{-\frac{\log(t+1)}{const}(l(Z_c(t-1)) - l(Z_c(t)))}, 1\right), \tag{9.14}$$

where $l(\cdot)$ denotes the likelihood (9.12) computed at Z_c, with all other parameters set at the current values. If $Z_c(t)$ does not satisfy (9.11), or it is not accepted, then $Z_c(t) = Z_c(t-1)$. After T steps, the outlier indicator matrix Z is updated, and set equal to the $Z_c(t)$, $t = 1, \ldots, T$, corresponding to the largest classification likelihood obtained. This finishes the S step. There are two tuning parameters at the S step of the `sclust` algorithm, T and *const*. The latter is used in (9.14) to control the acceptance probability. A large *const* will lead to a greedy stochastic optimizer, while a small *const* will allow some candidates leading to smaller likelihood to be accepted, in order to excape local optima for Z. Few pilot runs can be used to fine tune *const*. The other tuning parameter, T, shall be set large enough so that the stochastic optimizer is allowed to improve the current Z. A more efficient and less computationally intensive approach is to iterate the S step for $t \geq 1$ until a proposal is accepted, or a certain T has been reached. Hence, the S step may actually stop before the pre-specified number of iterations T.

The M step is very similar to the M step of the `tclust` algorithm described before. First of all, $\pi_c \propto \sum_i u_{ic}$, while for $c = 1, \ldots, k$ and $j = 1, \ldots, p$

$$\mu_{cj} = \frac{\sum_i z_{ij} x_{ij} u_{ic}}{\sum_i z_{ij} u_{ic}}. \tag{9.15}$$

An initial estimate of Σ_c is obtained by fixing

$$\Sigma_c = \frac{\sum_i u_{ic}(z_i(x_i - \mu_c))(z_i(x_i - \mu_c))^T}{\sum_i u_{ic} z_i z_i^T}, \tag{9.16}$$

Algorithm 9.3 Robust classification EM algorithm with component-wise outliers (`sclust`): CE and snipping step.

Classification E-step
for $i = 1, \ldots, n$ do
    if $\sum_j z_{ij} > 0$ then

$$\tilde{c}_i = \arg\max_c \pi_c \phi_{\sum_j z_{ij}} \left(x_{i(z_i)}, \mu_{c(z_i)}, \Sigma_{c(z_i)} \right)$$

        $u_{i\tilde{c}_i} = 1$
    end if
    for $c \neq \tilde{c}_i$ do
        $u_{ic} = 0$
    end for
end for
Snipping step
$Z_c(0) = Z$
for $t = 1, \ldots, T$ do
    Obtain $Z_c(t)$ by switching a zero and a one entry in $Z_c(t-1)$.
    if $Z_c(t)$ satisfies (9.10) and (9.11) **then**
        Accept $Z_c(t)$ with probability p_t as in (9.14), otherwise set $Z_c(t) = Z_c(t-1)$.
    else
        Set $Z_c(t) = Z_c(t-1)$.
    end if
    if $l(Z_c(t)) > l(Z)$ **then**
        $Z = Z_c(t)$
    end if
end for

---

and then modified in order to satisfy the constraint (9.4). To this end, $2kp+1$ function evaluations suffice to impose the eigenvalue constraint, exactly as in the `tclust` algorithm of Fritz *et al.* (2013). Note that the denominator of (9.15) is positive due to (9.10), and the denominator of (9.16) is positive due to (9.11).

If there are additional constraints on Z, these are directly imposed at the S step by discarding any $Z_c(t)$ which does not satisfy them.

The proposed algorithm is summarized as Algorithm 9.3 for the CE and snipping steps, and Algorithm 9.4 for the M step.

We conclude this section by noting that by fixing $k = 1$, a robust location and scatter estimator is obtained. Unlike the MCD and similar estimators, this estimator will be robust to component-wise outliers. The algorithm for computing this estimator, which we call `snipEM`, corresponds to a SM algorithm which alternates only the snipping and maximization steps. No restrictions on the condition number are needed when $k = 1$. By adding concentration steps

---

Algorithm 9.4 Robust classification EM algorithm with component-wise outliers (`sclust`): M step.

M-step
  for $c = 1, \ldots, k$ **do**
    for $j = 1, \ldots, p$ **do**

$$\mu_{cj} = \frac{\sum_i z_{ij} x_{ij} u_{ic}}{\sum_i z_{ij} u_{ic}}.$$

    end for

$$\Sigma_c = \frac{\sum_i u_{ic}(z_i(x_i - \mu_c))(z_i(x_i - \mu_c))^T}{\sum_i u_{ic} z_i z_i^T}$$

  Fix

$$\Sigma_c = V_c^T \Lambda_c V_c.$$

  end for
  if $\frac{\max_{jc} \lambda_{jc}}{\min_{jc} \lambda_{jc}} < const$ **then**
    Set $\hat{\theta}$ as the minimizer of

$$\sum_{c=1}^{k} \sum_{i=1}^{n} u_{ic} \sum_{j=1}^{d} \left(\log(\lambda_{jc}(\theta)) + \frac{\lambda_{jc}}{\lambda_{jc}(\theta)} \right)$$

    for $c = 1, \ldots, k$ **do**

$$\Sigma_c = V_c^T \Lambda_c(\theta) V_c,$$

    where $\Lambda_c = \text{diag}(\lambda_{jc}^m)$.
    end for
  end if

---

to perform trimming, one can obtain an estimator that is simultaneously robust to structural and component-wise outliers. The latter is implemented as function `stEM` in library `snipEM` but only for the case $k = 1$.

9.2.2   Properties

The theoretical properties of the `sclust` procedure as proposed are not investigated in Farcomeni (2014a). We develop them in the following.

First of all, it can be shown that the proposed `sclust` algorithm is convergent regardless of T. More formally

Theorem 9.1 *Fix $T \geq 1$ and const > 0, and denote with θ_j the estimate of θ obtained at the jth step of the CEM algorithm. We have that*

$$l(\theta_{j_1}) \leq l(\theta_{j_2}) \quad \forall \; j_1 \leq j_2. \tag{9.17}$$

Furthermore, there exists $J \in \mathcal{N}$ such that $l(\theta_{j_1}) = l(\theta_{j_2}) \quad \forall J \leq j_1 \leq j_2$, and that $\theta_{j_1} = \theta_{j_2} \quad \forall J \leq j_1 \leq j_2$.

Proof We first show that the classification likelihood is non decreasing in the sequence of solutions. The current parameter vector is as follows:

$$\theta_j = (\pi^{(j)}, \mu^{(j)}, \Sigma^{(j)}, Z^{(j)}, U^{(j)}).$$

The stochastic search is designed so that

$$l(\pi^{(j)}, \mu^{(j)}, \Sigma^{(j)}, Z^{(j+1)}, U^{(j)}) \geq l(\theta_j).$$

For a given Z, we assign observations to clusters using (9.13), which by construction maximizes $l(\pi^{(j)}, \mu^{(j)}, \Sigma^{(j)}, Z^{(j+1)}, U)$ with respect to U. Hence,

$$l(\pi^{(j)}, \mu^{(j)}, \Sigma^{(j)}, Z^{(j+1)}, U^{(j+1)}) \geq l(\theta_j).$$

At this point of the iteration we use a classical M step, which if the MLE is well defined leads to (9.17). See for instance Celeux and Govaert (1992). Note that the MLE is as a matter of fact well defined under modeling assumptions and restrictions, (9.11), and (9.4). The last two statements are a direct consequence of boundedness of the likelihood and of the properties of the MLE. To see the first, note that there is a finite number of partitions in k clusters and a finite number of snipping matrices Z. For each of them under (9.11), and assuming that clusters are non-empty, the likelihood is bounded given that the MLE is well defined. Concerning the last statement, for μ and Σ it does hold since the MLE is well defined. For U it does hold since $D_i(c) \neq D_i(d)$ when $c \neq d$ given that clusters are separated, and for π it holds as a consequence of this fact and the very definition of the E-step. For Z it holds with the same reasoning, as $l_c(\pi, \mu, \Sigma, Z_1, U) \neq l(\pi, \mu, \Sigma, Z_2, u)$ whenever $Z_1 \neq Z_2$ given that contribution to the complete likelihood are entry-specific.
□

Theorem 9.1 does not show that the proposed algorithm leads to a local optimum of the classification likelihood, but only that (i) it will stop and that (ii) the final estimates will correspond to an improved solution, that is, to a classification likelihood that is at least as large as that of the starting solution.

In order to guarantee that the algorithm is convergent to a local optimum of the likelihood, we need that the starting solution is in a suitable neighborhood of the optimum, or that the algorithm visits that neighborhood, and that T is large enough. Notably, in that case the final Z will be the global optimum conditional on the final estimates for the other parameters. Formally:

Theorem 9.2 *Let $T \to \infty$ and $const > 0$. Assume*

$$\theta^* = (\pi^*, \mu^*, \Sigma^*, Z^*, U^*)$$

is a local maximum for $l(\theta)$, with θ constrained by (9.11) and (9.4). It exists in a neighborhood of (π^*, μ^*, Σ^*) such that whenever the iterate $\theta^{(j)}$ is such that $(\pi^{(j)}, \mu^{(j)}, \Sigma^{(j)})$ falls in that neighborhood, the algorithm converges to θ^* at a linear rate.

Proof The proof follows along similar lines of Proposition 2 in Celeux and Govaert (1992). We just need to show that the algorithm can be seen as a grouped coordinate ascent method to optimize the classification likelihood. See also Bezdek *et al.* (1987), p. 473. More precisely, it suffices to show that we can describe the estimation algorithm as alternating two steps:

$$(Z^{(j+1)}, U^{(j+1)}) = \arg\max_{Z,U} l(\pi^{(j)}, \mu^{(j)}, \Sigma^{(j)}, Z, U) \qquad (9.18)$$

and

$$(\pi^{(j+1)}, \mu^{(j+1)}, \Sigma^{(j+1)}) = \arg\max_{\pi,\mu,\Sigma} l(\pi, \mu, \Sigma, Z^{(j+1)}, U^{(j+1)}) \qquad (9.19)$$

The second equation is exactly a consequence of the M-step, while the first results from (9.13) and the fact that for diverging T we achieve the global optimum for Z, conditionally on the other parameter values. This happens since there are a finite number, albeit large, of possible configurations of Z. Given (9.14), it can be shown that the rate of convergence of the stochastic optimizer is about $1/2$, and a function of the maximal difference in the classification likelihood when switching two entries in Z^* (Chakraborty and Chaudhury, 2008; Farcomeni, 2014b).
□

As far as robustness is concerned, Farcomeni (2014a) shows that the (universal) breakdown points for $\hat{\Sigma}$ are such that $\varepsilon^{(i)} \geq \varepsilon$ and $\varepsilon^{(c)} = \varepsilon + k/np$, in parallel with the results obtained in Ruwet *et al.* (2013) for trimmed k-means. The restricted breakdown points for $\bar{X}$ are also such that $\varepsilon^{(i)} \geq \varepsilon$ and $\varepsilon^{(c)} \geq \varepsilon$, which are the same bounds obtained for snipped k-means in Farcomeni (2014b).

In summary, if there is (only) structural contamination, `sclust` is as robust as `tclust`, which is based on trimming. If there is (also) component-wise contamination, `sclust` is more and more robust than `tclust` as the dimensionality grows.

9.3 Examples

9.3.1 Metallic oxide data

We begin by revisiting the metallic oxide data. To illustrate the advantages of using model based clustering as compared to k-means, we report once again the results of classical k-means procedure, trimmed k-means, and snipped k-means.

We use different trimming and snipping levels, whose results are shown in Table 9.1. The Rand indices are based on observations that are not trimmed by any of the procedures for fixed ε.

TABLE 9.1
Metallic oxide data. Comparison of the modified Rand index for different trimming/snipping levels. A comparison across different trimming levels is not reliable since the Rand index is computed on a different number of elements.

ε	tkmeans	skmeans	tclust	sclust
5%	17.2%	17.2%	24.0%	24.0%
7.5%	22.1%	26.0%	22.1%	26.0%
10%	20.0%	26.0%	27.9%	30.0%
12.5%	17.6%	19.1%	19.7%	40.8%
15%	17.6%	26.0%	25.9%	49.2%
17.5%	14.7%	26.0%	24.0%	38.9%

When $\varepsilon = 5\%$, the results of tclust and sclust coincide. There are two rows which must be trimmed, being structural outliers with extra low metal content. When $\varepsilon > 5\%$, snipping procedures are able to remove isolated entries in addition to the two structural outliers, better unveiling the true labels. It shall be noted that overall model based approaches perform much better than distance based ones. This is a clear indication that the assumption that clusters are spherical may be restrictive for the data at hand.

The different performance for different ε is not suggesting that the procedures are sensitive to the choice of ε, as the Rand index shall be interpreted and compared only row-wise and not column-wise in Table 9.1.

We now use the G statistic to choose the trimming and snipping levels. Once again we based our G statistic on the mean square distance between the centroid estimated with the current and a zero snipping or trimming level. We already reported that this choice leads to select $\varepsilon = 10\%$ for trimmed and snipped k-means. With robust model based approaches we select $\varepsilon = 5\%$ for tclust and $\varepsilon = 7.5\%$ for sclust. We do not report the actual G statistics but only mention that for ε larger than the levels chosen the G statistics are approximately constant, indicating that there is sensitivity to the choice of ε *only* when this is too low. For ε above a threshold, outliers are discarded. As usual we select exactly the threshold as snipping or trimming level to achieve the best possible efficency while performing robust estimation.

It shall be noted that it is not surprising that distance based methods, which are more restrictive, lead to slightly larger trimming/snipping levels. Simpler cluster shapes should make more observations look suspicious. Furthermore, it is not surprising that with sclust we have a slightly larger ε than with tclust. An increase in the trimming level would lead to flag entire observations with many clean entries, and this may not be efficient for centroid estimation. On the other hand, an increase in the snipping level may lead to flag isolated entries, which may improve the centroid estimate sensibly.

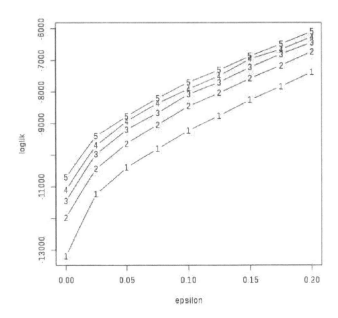

FIGURE 9.1
Water treatment plant data. Maximum of the likelihood for different k and ε.

9.3.2 Water treatment plant data

We now turn to the daily measurements of sensors in a urban waste water treatment plant. We mostly focus on `sclust` for the time being.

We begin by using the trimmed likelihood curve methodology of García-Escudero *et al.* (2011) in order to select the number of clusters. From Figure 9.1 it can be seen that the optimum of the objective function is rather close for $k = 3, 4, 5$ when ε is above a threshold. On the other hand, there is a clear advantage in passing from $k = 1$ to $k = 2$ and from $k = 2$ to $k = 3$. On the basis of these considerations we fix $k = 3$.

We now choose ε. To do so, we compute the G statistic based on the unstandardized Euclidean distance among centroids, that is,

$$G(\varepsilon) = \sum_{jc} (\mu_{jc}(\varepsilon) - \mu_{jc}(0))^2, \qquad (9.20)$$

for $k = 3$ and different values of ε. A plot of the G statistics versus ε is given in Figure 9.2. The plot suggests that the estimates are fairly stable for $\varepsilon \geq 0.05$, so that we end up setting $\varepsilon = 0.05$. Note that this suggestion is confirmed by the likelihood curve in Figure 9.1, as for $\varepsilon \geq 0.05$ the curves for $k = 3, 4, 5$ are

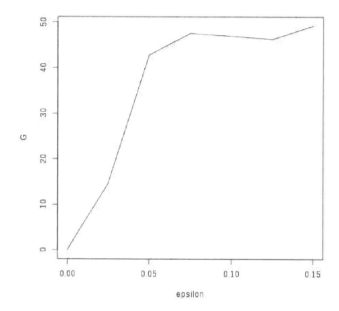

FIGURE 9.2
Water treatment plant data. G statistics when $k = 3$.

rather close. Figure 9.2 gives also substantial evidence that there are outliers in the data, so that classical methods may be biased.

We therefore use `sclust` with $k = 3$ and $\varepsilon = 0.05$. For reasons of space we do not report the parameter estimates, but only a summary of them. These are obtained by standardizing the clean entries in the data matrix, and averaging groups of measurements by location. Note that we could not do this operation on the original data matrix due to the possibility of outliers (which are indeed estimated as about 5% of the entries), while the operation is safe when restricted to entries not flagged by the robust algorithm. We report only on the input, first and second settler, and output in Table 9.2. We can safely conclude that the cluster analysis has identified groups of days with about average (group 1, about 40% of the observations), low (group 2, 25%) and heavy (group 3, the remaining 35%) loads to the plant. The groups are well separated as testified by the Fisher's F statistic, which corresponds to 22.3. An important outcome of this analysis involves the distribution of the number of consecutive days in cluster 3, which identifies days of heavy duty. This information shall be used for planning of plant use and prevention of faults. With our algorithm we can estimate an average of 1.6 consecutive days

TABLE 9.2
Water treatment plant data. Average of standardized clean measurements by
location and cluster as identified by `sclust`.

	Cluster 1	Cluster 2	Cluster 3
Input	-0.062	-0.385	0.327
First Settler	-0.024	-0.537	0.393
Second Settler	0.089	-0.640	0.388
Output	0.034	-0.333	0.154

TABLE 9.3
Water treatment plant data. Average of standardized clean measurements by
location and cluster as identified by `tclust`.

	Cluster 1	Cluster 2	Cluster 3
Input	-0.072	-0.043	0.066
First Settler	0.013	-0.138	0.013
Second Settler	0.084	-0.114	0.096
Output	0.027	0.029	0.00

in state 3, standard deviation 0.97, with a probability of persistence of two or
more consecutive days in state 3 of 37%. On the other hand, if we use classical
Gaussian mixture models, we obtain an average consecutive number of days
in state 3 of 1.47, standard deviation 0.88, with a probability of persistence
for two or more days only 25%. Consequently, Gaussian mixture models that
ignore outlying entries may lead to significantly underestimate the activity
and load of the plant.

We now briefly compare our previous results, which are based on `sclust`,
with those of `tclust`. We select $\varepsilon = 7.5\%$ using the G statistic criterion, and
three groups with 30%, 27% and 43% of the clean observations, respectively.
A total of 29 days are trimmed. In Table 9.3 we report on the estimated
centroids as before, where it can be seen that there may be some bias (due
to the presence of structural outliers). This is testified for instance by the
counter intuitive situation for which in cluster 2 we have much lower inputs
than cluster 3, but larger output.

9.1 R Illustration (Robust clustering metallic oxide data) *The following is a sample code to implement* `tclust` *and* `sclust` *on the Metallic Oxide
data. For* `sclust`, *the outcome is based on a single initial solution. Note that
the problem may become ill-conditioned if the initial solution is far from the
optimal one. In real applications around 50 initial solutions, along the lines*

of tclust, *should be tried and compared with respect to the likelihood at* *convergence.*

```
> library(tclust)
> source("rand.index.r")
> library(snipEM)
>
> eps=c(0.05,0.075,0.1,0.15,0.175)
>
> res=matrix(NA,length(eps),7)
> res[,1]=eps
>
> gstat=matrix(NA,length(eps),2)
> g0=kmeans(fell[,-1],2)$cent
> g0=g0[order(g0[,1]),]
>
> i=1
> for(ep in eps) {
+ tk=tclust(fell[,-1],2,alpha=ep)
+ V=matrix(1,31,8)
+ V[24:25,]=0
+ if(ep>0.05) {
+ u=pmin(mahalanobis(fell[,-1],tk$cent[,1],
+ tk$cov[,,1]),
+ mahalanobis(fell[,-1],tk$cent[,2],tk$cov[,,2]))
+ V[abs(u)>=quantile(unlist(abs(u)),1-ep)]=0}
+ R=apply(sapply(1:2, function(j)
+ mahalanobis(fell[,-1],
+ tk$cent[,j],tk$cov[,,j])),1,which.min)
```

In the code above the initial Z (V *in the code) is chosen so that rows 24* *and 25 are structural outliers. Additional outliers are then initialized as those* *exceeding the* $1 - \varepsilon$ *quantile of the Mahalanobis distances from the closest* *centroid. The initial cluster labels are assigned accordingly.*

We then proceed within the loop to perform sclust, *and evaluate through* *computation of the G statistics and Rand index on rows not trimmed by any* *of the procedures.*

```
+ sk=sclust(as.matrix(fell[,-1]),2,V,R)
+ gstat[i,1]=mean((t(tk$cent[,order(tk$cent[1,])])
+ -g0)^2)
+ gstat[i,2]=mean((sk$m[order(sk$m[,1]),]-g0)^2)
+ u1=tk$clust!=0
+ u2=apply(sk$V,1,sum)>0
+ u=u1 & u2
+ res[i,-1]=c(rand.index(unclass(fell[u,1]),
```

```
+ tk$clust[u]),
+ rand.index(unclass(fell[u1,1]),tk$clust[u1]),
+ sum(tk$clust==0),rand.index(unclass(fell[u,1]),
+ sk$R[u]),
+ rand.index(unclass(fell[u2,1]),sk$R[u2]),
+ sum(apply(sk$V,1,sum)==0))
+ i=i+1}
```

10

Double Clustering

There are many applications in which one may want to cluster both rows and columns. For instance, in DNA microarray analysis the observed expression of p genes on n slides is recorded, and while clustering the genes is of primary interest, clustering the slides leads to identification of groups among the patients and is useful as well. Other applications include marketing (for instance, clustering customers and goods), biology, psychology, sociology. Following arguments that date back at least to Fisher (1969) and Hartigan (1972), it can be argued that the most appropriate route in these cases is to perform *simultaneous* clustering of rows and columns. This is called *biclustering* or *double clustering* in the statistical literature.

Reviews of double clustering can be found in Van Mechelen *et al.* (2004) and Madeira and Oliveira (2004), for instance. While there are now many competing strategies for performing double clustering in many different situations, to the best of our knowledge the only formally robust approach is that of Farcomeni (2009b). Robustness in double clustering is crucial, as procedures are often used in connection with large data matrices in which contamination may occur row-wise, column-wise and component-wise.

It is important to underline that clustering variables is conceptually similar but technically different from dimension reduction, which has been discussed in the first part of this book. Double clustering can as a matter of fact be seen as simultaneous dimension and sample reduction, but many formal properties of dimension reduction (e.g., that the variance of the new variables is maximized) are lost. The "packing" matrix of eigenvectors V is here a binary matrix. The scores with double clustering are obtained as a simple average, rather than as a (signed) weighted average of the old variables. More precisely, in double clustering the loadings are either zero or one. In order to underline the parallel with dimension reduction methods, we will denote with V the matrix of column clusters; while we still denote with U the matrix of row clusters.

There are also procedures that simultaneously perform clustering and dimension reduction, e.g., Tipping and Bishop (1999), McNicholas and Murphy (2008), Hou *et al.* (2013). These techniques have a somewhat different aim, and are beyond our scope.

In this chapter we will review symmetric methods in the definition of Section 2.2.3 of Van Mechelen *et al.* (2004): the same variant is obtained by applying the algorithm to X^T instead of X. For an asymmetric robust double clustering method see for instance Algorithm 2 in Farcomeni (2009b).

We begin by reviewing the classical "double k-means" procedure, and then two robust versions: one is based on trimming and was proposed in Farcomeni (2009b), the second is based on snipping and is an original contribution we give in this book.

10.1   Double k-means

Double k-means proceeds by simultaneously clustering rows and columns of a data matrix. Therefore, even if its name does not suggest it, there are *two* discrete parameters to be specified. One, k_1, is the number of row clusters; the other, k_2, denotes the number of column clusters.

The objective function connected with non-robust double k-means can be expressed in matrix form as (see, e.g., Van Mechelen *et al.* (2004)):

$$||X - U\bar{X}V^T||; \tag{10.1}$$

where differently from the previous chapters both U and V are binary, and their rows sum to 1.

The outcome of a double clustering procedure can be interpreted from different point of views.

The centroid matrix has dimension $k_1 \times k_2$. In this sense, double k-means can be seen as a stronger form of data reduction as soon as $k_2 < p$. Recall in fact that the centroid matrix has dimension $k \times p$ in k-means. The row cluster profiles in the rows of $\bar{X}$ in double clustering do not summarize a location with respect to each variable, but only with respect to each column cluster. The row cluster membership matrix U has dimension $n \times k_1$, and gives cluster membership for the observations. Similarly and symmetrically, the column cluster membership matrix V is $p \times k_2$; and gives a somewhat new information, regarding grouping of variables. A cluster of variables can be seen as a cluster of measurements which have very close means within each row cluster. For instance, in genomic applications we put slides on the rows and genes on the columns of the data matrix. A column cluster indicates very similar levels of expression for a group of genes *within each row cluster*. This is in parallel with the interpretation of row clusters, but usually the interpretation is not intuitive at first. Column cluster profiles are obtained from the columns of $\bar{X}$.

The conventional double k-means algorithm is given as Algorithm 10.1. It can be seen as a direct generalization of the conventional k-means algorithm, in which the row and column memberships are updated with similar strategies. It can be easily shown that at each iteration of Algorithm 10.1 the

---

Algorithm 10.1 Double k-means

Compute the distances for row profiles

for $i = 1, \ldots, n$ **do**

Let $d_{ir}^2 = \sum_{j=1}^{p} \sum_{c=1}^{k_2} v_{jc}(x_{ij} - \bar{x}_{rc})^2$, $r = 1, \ldots, k_1$.

Let $r_i = \text{argmin}_r d_{ir}^2$.

end for

Update the row memberships

for $i = 1, \ldots, n$ **do**

Let $d_{ir}^2 = \sum_c \sum_{j=1}^{p} v_{jc}(x_{ij} - \bar{x}_{rc})^2$, $r = 1, \ldots, k_1$

Let $r_i = \text{argmin}_r d_{ir}^2$.

Set $u_{ir_i} = 1$. All the other elements of the ith row of U are set to 0.

end for

Compute the distances for column profiles

for $j = 1, \ldots, p$ **do**

Let $d_{jc}^2 = \sum_{i=1}^{n} \sum_{r=1}^{k_1} u_{ir}(x_{ij} - \bar{x}_{rc})^2$, $c = 1, \ldots, k_2$.

Let $c_j = \text{argmin}_c d_{jc}^2$.

end for

Update the column memberships

for $j = 1, \ldots, p$ **do**

Let $d_{jc}^2 = \sum_{i=1}^{n} \sum_r u_{ir}(x_{ij} - \bar{x}_{rc})^2$, $c = 1, \ldots, k_2$.

Let $c_j = \text{argmin}_c d_{jc}^2$.

Set $v_{jr_j} = 1$. All the other elements of the jth row of V are set to 0.

end for

Estimate the centroid matrix

$\bar{X} = (U^T U)^{-1} U^T X V (V^T V)^{-1}$.

---

objective function (10.1) decreases, and at convergence we have at least a local minimum.

The most innovative feature of Algorithm 10.1 is that distances are computed twice. First, we compute n (Euclidean) distances summarizing the distance of the rows from their closest (row) centroid. It shall be noted that in the computation of d_{ir}^2 we use the column membership matrix V. In this way, the distance of the jth measurement for the ith observation is computed with respect to the profile of the cluster of the jth variable. To be more clear, for each j we have a sum of k_2 elements, but only one of them will be positive. After we have obtained the distances for the row profile, the algorithm proceeds similarly to the conventional k-means algorithm in order to update the row memberships. Then, we compute p Euclidean distances summarizing the distance of the columns from their closest centroid, and proceed updating the column memberships. Finally, a least squares estimate of the centroid matrix is readily obtained.

10.2   Trimmed double k-means

In this section our model is still (10.1), the only difference being that we allow for the presence of (row and column) structural outliers. It results in a double trimmed k-means procedure, which is very similar to Algorithm 1 in Farcomeni (2009b). Given that we combined two trimmed k-means procedures, we specify two trimming levels, ε_1 and ε_2, for the rows and columns respectively. The trimmed procedure is obtained by constraining $n\varepsilon_1$ rows of U to sum exactly to zero and similarly for $p\varepsilon_2$ rows of V. Note that when $p << n$, as often happens in small to moderate scale applications, it may be sensible to set $\varepsilon_2 = 0$, obtaining conventional k-means for the columns.

Even if the setting of double clustering is slightly different, we can adopt a spurious outlier model that is similar to that of trimmed k-means. We assume the $(ij)^{th}$ *clean* entry of the data arises from an elliptical distribution centered in

$$\sum_r \sum_c u_{ir} v_{jc} \bar{x}_{rc}.$$

If the $(ij)^{th}$ entry belongs to a column *or* row outlier, then it is assumed to arise from g_{ij}, an unknown and unspecified statistical law that generates the outliers in a general position.

As usual we assume affine independence and separation for g_{ij}. Consequently, we can ignore the contribution of nonregular observations and minimize an objective function of the trimmed k-means type:

$$\min_{\bar{X},U,V,z_1,z_2} \sum_r \sum_c u_{ir} v_{jc} z_{1i} z_{2j} (x_{ij} - \bar{x}_{rc})^2, \qquad (10.2)$$

where z_1 is a binary vector of length n such that $\sum_i z_{1i} = n(1 - \varepsilon_1)$, and z_2 is a binary vector of length p such that $\sum_j z_{2j} = p(1 - \varepsilon_2)$.

The robust double clustering model can be fit by alternating concentration steps similarly to the trimmed k-means procedure. Given current estimates for U, $\bar{X}$ and V, the general iteration is given in Algorithm 10.2.

It can be seen that Algorithm 10.2 is very similar to Algorithm 10.1, where we include concentration steps in order to flag the most extreme distances.

In order to choose a starting solution, a simple possibility is to initialize U and z_1 using trimmed k_1-means and then V and z_2 with a trimmed k_2-means on the transpose of the data matrix. An initialization approach specifically tailored for double clustering, but which does not take into account the possibility of contamination, is suggested in Cho *et al.* (2004).

Algorithm 10.2 Trimmed double k-means

Compute the distances for row profiles
for $i = 1, \ldots, n$ do
$\quad$ Let $d_{ir}^2 = \sum_{c=1}^{k_2} \sum_{j=1}^{p} v_{jc}(x_{ij} - \bar{x}_{rc})^2, \; r = 1, \ldots, k_1$.
end for
Concentration step for the rows
Sort the row distances to the closest centroid as

$$d_{i_1 r_{i_1}}^2 \leq d_{i_2 r_{i_2}}^2 \leq \cdots \leq d_{i_n r_{i_n}}^2$$

Set $z_{1 i_1} = z_{1 i_2} = \cdots = z_{1 i_{n(1-\varepsilon_1)}} = 1$, all other elements of z_1 are set to zero.
Update the row memberships
for $i = 1, \ldots, n$ do
$\quad u_i = 0$
$\quad$ if $z_{1i} = 1$ then
$\quad\quad$ Let $d_{ir}^2 = \sum_c \sum_{j=1}^{p} v_{jc}(x_{ij} - \bar{x}_{rc})^2, \; r = 1, \ldots, k_1$
$\quad\quad$ Let $r_i = \operatorname{argmin}_r d_{ir}^2$. Set $u_{i r_i} = 1$.
$\quad$ end if
end for
Compute the distances for column profiles
for $j = 1, \ldots, p$ do
$\quad$ Let $d_{jc}^2 = \sum_{r=1}^{k_1} \sum_{i=1}^{n} u_{ir}(x_{ij} - \bar{x}_{rc})^2, \; c = 1, \ldots, k_2$.
end for
Concentration step for the columns
Sort the column distances to the closest centroid as

$$d_{j_1 c_{j_1}}^2 \leq d_{j_2 c_{j_2}}^2 \leq \cdots \leq d_{j_p c_{j_p}}^2$$

Set $z_{2 j_1} = z_{2 j_2} = \cdots = z_{2 j_{p(1-\varepsilon_2)}} = 1$, all other elements of z_1 are set to zero.
Update the column memberships
for $j = 1, \ldots, p$ do
$\quad v_j = 0$
$\quad$ if $z_{2j} = 1$ then
$\quad\quad$ Let $d_{jc}^2 = \sum_r \sum_{i=1}^{n} u_{ir}(x_{ij} - \bar{x}_{rc})^2, \; c = 1, \ldots, k_2$.
$\quad\quad$ Let $c_j = \operatorname{argmin}_c d_{jc}^2$. Set $v_{j r_j} = 1$.
$\quad$ end if
end for
Estimate the centroid matrix
$\bar{X} = (U^T U)^{-1} U^T X V (V^T V)^{-1}$.

10.3 Snipped double k-means

Double trimmed k-means is as expected robust with respect to structural contamination. It may be inefficient with component-wise contamination. On the one hand, it may lead to discard valuable information. If in fact one single dimension for some (row or column) vector is contaminated, the entire vector may (and hopefully will) be marked as outlying. The loss of information may not be negligible. On the other hand, as illustrated before when introducing snipped k-means, the contaminated entries may be so scattered that there is at least one contaminated entry per row or per column. In this case, with trimmed double k-means ε_1 or ε_2 would have to be set to an unfeasible level.

In this section we make use of a unique identifier of contaminated entries, a binary $n \times p$ matrix Z, with zeros identifying contaminated entries as usual. Further, a unique snipping level ε is set. As with snipped k-means, the Z matrix is constrained so that $\sum_{ij} z_{ij} = np(1 - \varepsilon)$.

The objective function for double snipped k-means is

$$\min_{\bar{X},U,V,Z} \left(\sum_{i=1}^{n} \sum_{j=1}^{p} \sum_{r=1}^{k_1} \sum_{c=1}^{k_2} z_{ij} u_{ir} v_{jc} (x_{ij} - \bar{x}_{rc})^2 \right). \tag{10.3}$$

Note that the regular observations contribute with all their dimensions to the objective function, while as usual snipped observations contribute only with a subset. Furthermore, among the $k_1 k_2$ distances of x_{ij} to each element of the centroid matrix $\bar{X}$, only the one corresponding to the combination of row and column clusters identified by $u_{ir} v_{jc}$ contributes to the objective function. The proposed double snipped k-means is in the spirit of Algorithm 2 in Farcomeni (2009b), but rather different from what is in that paper.

Once again we could use a concentration step to update Z, but in order to avoid local optima we use a stochastic optimization approach. The parameter Z is therefore updated through a stochastic snipping step. A sequence of T solutions is built by switching a zero and a one entry in Z uniformly at random, and accepting the proposal with probability

$$p_t = \min\left(e^{-\frac{\log(t+1)}{const}(l(Z_c(t)) - l(Z_c(t-1)))}, 1 \right), \tag{10.4}$$

where *const* and T can be tuned as usual. Given current estimates for U, $\bar{X}$, V and Z the general iteration for double snipped k-means is reported in Algorithm 10.3.

10.4 Robustness properties

Farcomeni (2009b) defines the *bi-mode breakdown point* for a centroid matrix

---

Algorithm 10.3 Snipped double k-means

Compute the distances for row profiles
for $i = 1, \ldots, n$ **do**

Let $d_{ir}^2 = \sum_{c=1}^{k_2} \sum_{j=1}^{p} z_{ij} v_{jc} (x_{ij} - \bar{x}_{rc})^2$, $r = 1, \ldots, k_1$.

end for
Update the row memberships
for $i = 1, \ldots, n$ **do**

$u_i = 0$
if $\sum_j z_{ij} > 0$ **then**

Let $d_{ir}^2 = \sum_c \sum_{j=1}^{p} z_{ij} v_{jc} (x_{ij} - \bar{x}_{rc})^2$, $r = 1, \ldots, k_1$

Let $r_i = \operatorname{argmin}_r d_{ir}^2$; set $u_{ir_i} = 1$
end if

end for
Compute the distances for column profiles
for $j = 1, \ldots, p$ **do**

Let $d_{jc}^2 = \sum_{r=1}^{k_1} \sum_{i=1}^{n} z_{ij} u_{ir} (x_{ij} - \bar{x}_{rc})^2$, $c = 1, \ldots, k_2$.

end for
Update the column memberships
for $j = 1, \ldots, p$ **do**

$v_j = 0$
if $\sum_i z_{ij} > 0$ **then**

Let $d_{jc}^2 = \sum_r \sum_{i=1}^{n} z_{ij} u_{ir} (x_{ij} - \bar{x}_{rc})^2$, $c = 1, \ldots, k_2$.

Let $c_j = \operatorname{argmin}_c d_{jc}^2$; set $v_{jc_j} = 1$.
end if

end for
Update Z
$Z_c(0) = Z$
for $t = 1, \ldots, T$ **do**

Obtain $Z_c(t)$ by switching a 0 and 1 entry in $Z_c(t-1)$.
Accept $Z_c(t)$ with probability p_t as in (10.4)
if $l(Z_c(t)) < l(Z)$ **then**

$Z = Z_c(t)$
end if

end for
Estimate the centroid matrix
For $r = 1, \ldots, k_1$; $c = 1, \ldots, k_2$ estimate $\bar{X}$ as

$$\bar{x}_{rc} = \frac{\sum_{i=1}^{n} \sum_{j=1}^{p} z_{ij} u_{ir} v_{jc} x_{ij}}{\sum_{i=1}^{n} \sum_{j=1}^{p} z_{ij}}$$

---

as

$$\varepsilon^{bi-mode} = \min\left(\frac{1}{n}\max\{M_1 : \sup_{X_{M_1,0}} ||\bar{X}(X) - \bar{X}(X_{M_1,0})|| \in K\},\right.$$

$$\left.\frac{1}{p}\max\{M_2 : \sup_{X_{0,M_2}} ||\bar{X}(X) - \bar{X}(X_{0,M_2})|| \in K\}\right),$$

where $X_{r,c}$ is the data matrix in which r rows and c columns are replaced by arbitrary values. The bi-mode breakdown point is defined as the minimum between the individual breakdown point obtained when contaminating only the rows, and the one obtained when contaminating only the columns.

It is shown in Farcomeni (2009b) that for classical double k-means, $\varepsilon^c < 1/np$, and $\varepsilon^{bi-mode} < 1/\max(n,p)$. Both breakdown values are then infinitesimal. In Farcomeni (2009b) is also shown that for trimmed double k-means

$$\varepsilon^c < \frac{n\varepsilon_1 + p\varepsilon_2 + 1}{np},$$

and

$$\varepsilon^{bi-mode} < \frac{n\varepsilon_1 + p\varepsilon_2 + 1}{\max(n,p)}.$$

Consequently, trimmed double k-means *could* be formally robust, but we only have an upper bound for the breakdown point so we do not know for sure.

Formal lower bounds of the breakdown points of trimmed double k-means and snipped double k-means have not been yet studied and are grounds for further research. We speculate that with restriction to a class of well-clustered data sets it can be shown that the restricted breakdown points correspond to the upper bounds above for trimmed double k-means. Additionally we speculate that after restriction it should be possible to prove that for snipped double k-means $\varepsilon^c = \varepsilon^{bi-mode} = \varepsilon$ asymptotically.

10.1 Example (Robust double clustering G8 data) *We here develop an example about the macroeconomic performances of the most industrialized countries. For these data the common choice for the number of groups is $k_1 = 3$ and $k_2 = 2$.*

With the G statistic as originally defined in Farcomeni (2009b) and trimmed double k-means we end up setting $\varepsilon_1 = 1/8$ and $\varepsilon_2 = 0$. There is one single row outlier, which is identified by the algorithm as being Italy. The final row partition is: $\{GER, JAP, USA, CAN\}, \{SPA\}, \{FRA, GBR\}$. while the column partition is: $\{GDP, DEF, DEB, TRB\}, \{INF, INT, UNE\}$. Note that the low sample size makes it very likely that singletons are obtained, and in fact Spain is also an outlier given it is the only member of its cluster.

If we use snipped double k-means, the automatic choice based on the G statistic leads to set $\varepsilon = 2/56 = 0.0357$. Two entries must be snipped. After 50 random initial starting solutions, these are identified as USA's GDP, and Italy's debt. Italy is once again identified as being an outlier, but only for the variable DEB. For the other variables, Italy is not considered as an outlier, and the row-partition is the same as before, where Italy is classified in the same group as Spain. We can conclude that in this data set there is some evidence of Italy being an outlier, possibly because of an exceptionally high public debt. Further, it shall be noted that USA's GDP is identified as outlying since it is exceptionally high even after standardization.

10.2 R Illustration (Robust double clustering G8 Data) *For the G8 example, data are stored in a data frame* **g7**. *A multistart strategy based on 50 initial solutions is implemented in order to increase the likelihood of obtaining the global optimum.*

Snipped and Trimmed double k-means were implemented in functions which we provide with the accompanying material, and can be loaded in **R** *with*

```
> source("snipdkm.r")
> source("trimdkm.r")
```

A sample code for computing trimmed double k-means follows:

```
> X=scale(g7[,-1])
>
> tr=try(trimdkm(X,3,2,1,0),silent=TRUE)
> while(inherits(tr,"try-error")) {
+ tr=try(trimdkm(X,3,2,nr,nc),silent=TRUE)}
>
> for(i in 2:50) {
+ tr2=try(trimdkm(X,3,2,1,0),silent=TRUE)
+ if(!inherits(tr2,"try-error") && tr2$obj<tr$obj) {
+ tr=tr2}}
```

We now perform snipped double k-means and compute the G statistics for nine different contamination levels:

```
> gstatsn=rep(NA,9)
>
> for(nr in 1:9) {
+
+ Z=matrix(1,nrow(X),ncol(X))
+ Z[sample(nrow(X)*ncol(X),nr)]=0
+ sn=try(snipdkm(X,3,2,Z))
+ while(inherits(sn,"try-error"))
+ {sn=try(snipdkm(X,3,2,Z),silent=TRUE)}
>
> ## multistart
> for(i in 2:50) {
```

```
+ Z=matrix(1,nrow(X),ncol(X))
+ Z[sample(nrow(X)*ncol(X),nr)]=0
+ sn2=try(snipdkm(X,3,2,Z),silent=TRUE)
+
+ if(!inherits(sn2,"try-error") && sn2$obj<sn$obj) {
+ sn=sn2}}
+
+ gstatsn[nr]=sn2$obj}
```

11

Discriminant Analysis

The perspective on sample reduction in this chapter is rather different from that of the previous ones. Up to this point in the book we have assumed we have no measures of data labels, groups or strata. We have reduced the sample based on unsupervised procedures which obtained as many groups as were requested. In this chapter the number of groups k is known. In addition, we also have recorded without uncertainty the group label for each observation. Consequently, we have ready a matrix U with cluster labels which can be used for sample reduction. The centroid matrix $\bar{X}$ is readily available and does not need to be estimated, and similarly for any other statistics which may be stratified by group. Discriminant analysis is concerned with a slightly different objective than cluster analysis: that of *prediction* or *classification*, that is, assigning additional observations to one of the k groups. Before we proceed, let us underline that there are many methods for classification (see for instance Hastie *et al.* (2009)), where a review is beyond the scopes of this book. Discriminant analysis is only one of the possible options, which nicely fits into the framework of the book because of many similarities with topics like multivariate estimation of location and scatter, and cluster analysis. It shall be intended as an introduction to the topic and to the area of robust classification: discriminant analysis is limited by the fact that all predictors must be continuous, and that a parametric Gaussian assumption should be formulated possibly after transformation. Many classification methods on the other hand work well also with a mix of continuous (without parametric assumptions) and categorical measurements.

11.1 Classical discriminant analysis

This section is used to briefly introduce linear and quadratic discriminant analysis in the classical non-robust framework. The setting is rather similar to that of the previous chapters, where we assume that observations within a certain group arise from a Gaussian distribution with mean μ_c and covariance Σ_c.

Heteroscedastic or *quadratic* discriminant analysis uses little further assumptions, while *linear* discriminant analysis further assumes homogeneity of the covariance matrices, in that $\Sigma_c = \Sigma$ for $c = 1, \ldots, k$.

We have recorded information regarding group labels, that is, we have recorded the group of each observation. The total number of groups, k, is known. Discriminant analysis is a *supervised* classification method. There is no need to estimate groups as with unsupervised methods.

The two main reasons behind the analysis are (i) parameter estimation for interpretation and sample reduction and (ii) obtaining prediction rules for classification of future observations. The first task is rather straightforward and can be easily accomplished using the sample mean and covariance (or the pooled covariance under assumptions of homogeneity), stratified by group. The second task is the result of discriminant analysis. A detailed review can be found in Lachenbruch (1975), and more recently in McLachlan (1992) or Huberty (1994).

As far as sample reduction is concerned, one can estimate centroids as usual. The main difference with unsupervised approaches is that the binary cluster assignment matrix U is known, and shall not be estimated. Hence one can simply proceed by estimating, for $c = 1, \ldots, k$,

$$\hat{\mu}_c = \frac{\sum_{i=1}^n u_{ic} x_i}{\sum_{i=1}^n u_{ic}}; \tag{11.1}$$

and

$$\hat{\Sigma}_c = \frac{\sum_{i=1}^n u_{ic}(x_i - \hat{\mu}_c)(x_i - \hat{\mu}_c)'}{\sum_{i=1}^n u_{ic}}. \tag{11.2}$$

If there are homogeneity assumptions, the pooled covariance shall be obtained as follows:

$$\hat{\Sigma} = \frac{\sum_{c=1}^k \sum_{i=1}^n u_{ic}(x_i - \hat{\mu}_c)(x_i - \hat{\mu}_c)'}{n}. \tag{11.3}$$

For simplicity we now focus on *linear* discriminant analysis, that is, we assume homoscedasticity and use (11.3). We detail below what happens when relaxing this assumption. After estimation, a very popular approach to classification is to assign the new observation to the closest centroid as measured by the Mahalanobis distance. This rule is optimal if the groups are believed to be balanced *a priori*. The prior probabilities π_j, $j = 1, \ldots, k$ can be used to penalize the Mahalanobis distances, obtaining the Bayes discriminant rule which assigns the new observation x_i to the group which corresponds to the minimizer of

$$(x_i - \hat{\mu}_c)'\hat{\Sigma}^{-1}(x_i - \hat{\mu}_c) - 2\log\pi_c \tag{11.4}$$

over $c = 1, \ldots, k$. If prior information on the group sizes are not available, they can be easily estimated from the data as

$$\hat{\pi}_c = \frac{\sum_{i=1}^n u_{ic}}{n}.$$

The Bayes discriminant rule penalizes smaller clusters, so that new observations are less likely to be assigned to groups with small prior probabilities; and it can be shown to minimize the total probability of misclassification (Johnson and Wichern, 2002). It can be noted that the quadratic term $x_i'\hat{\Sigma}^{-1}x_i$ in (11.4) is constant over c, hence it can be omitted. As a consequence, (11.4) is equivalent to the *linear* (in x_i) expression

$$\hat{\mu}_c'\hat{\Sigma}^{-1}\hat{\mu}_c - 2x_i\hat{\Sigma}^{-1}\mu_c - 2\log\pi_c.$$

This is one of the reasons why homoscedastic discriminant analysis is referred to as linear discriminant analysis.

Under heteroscedasticity, the Bayes rule must also take into account the different covariance matrices, and the group of the ith new observation is the minimizer of

$$\log|\Sigma_c| + (x_i - \hat{\mu}_c)'\hat{\Sigma}_c^{-1}(x_i - \hat{\mu}_c) - 2\log\pi_c \qquad (11.5)$$

for $c = 1, \ldots, k$.

There are actually many important ideas in supervised classification which we can only briefly mention, and refer the reader to the detailed book Hastie *et al.* (2009).

An important one is related to evaluation of the performance of the classifier. When prediction is the target of the analysis, there actually often are many different classifiers available. The choice of the best classifier depends on its performance on the data at hand, and there is no uniformly better method. Here for instance we may want to choose between a simpler linear and a more complex quadratic classifier. First of all, we can blindly evaluate the performance of the classifier by estimating the conditional and marginal probabilities that an observation is misclassified. The first is defined, for $c = 1, \ldots, k$, as

$$\Pr(\hat{U}_{ic} = 0 \mid U_{ic} = 1), \qquad (11.6)$$

while the second as

$$\Pr(\bigcup_{c=1}^{k} \hat{U}_{ic} = 0 \bigcap U_{ic} = 1) = \qquad (11.7)$$

$$\sum_{c=1}^{k}\Pr(\hat{U}_{ic} = 0 \mid U_{ic} = 1)\Pr(U_{ic} = 1).$$

When looking at (11.6) and its weighted average (11.7), one must realize that the error probabilities are nonzero even for observed subjects. This is due to the fact that the model, as always, does not fit the data perfectly and therefore is not able to predict the correct labels even for the data used for estimation. Data used for estimation are often called *training set* in the supervised classification context. The k conditional error probabilities give an idea of what is the expected error rate for each class. It often happens that classes with smaller π_c, larger $|\Sigma_c|$, or with small separation (e.g., small $||\mu_c - \mu_d||$ for some

c and d) are more difficult to correctly predict. The marginal conditional error probability give instead an idea of the expected error rate regardless of the true class. The probabilities above are usually simply estiamated through their empirical frequencies observed, by comparing the known and estimated labels.

Another important issue in regards to the fact that the error probabilities estimated on the training data may be overly optimistic, that is, the actual error rates on new data may be slightly larger. This discrepancy is more and more evident as the model becomes more complex. Suppose we manage to specify an extremely complex model fitting perfectly the training set. It is intuitive that one cannot expect this model to perfectly predict the new data. There are many possibilities for obtaining an honest estimate of the error probabilities for the new data. A simple one when n is slightly large is by splitting the data in a training set, usually made of 50% to 80% of the available observations, and a *test set* which is put aside. The choice of which observations are included in the test set shall of course be performed at random, possibly uniformly at random if there are no reasons to do this differently. In our example we will actually never select the test set uniformly at random. Sometimes it is better in fact to perform stratified sampling, which selects a possibly unbalanced number of observations uniformly at random within each group. This is particularly relevant when there are very small groups, so that misclassification estimates are also based on the performance of the classifier in small groups (which are often of great interest). After the model is estimated, predictions are obtained for the test set and these predictions are compared with the available labels to obtain estimates for (11.6) and (11.7).

There are limitations of data splitting in certain cases. When n is small it may not be desireable to set aside half of the sample for error estimation, as this would significantly increase the standard errors. Additionally, 10-20% of the sample may be only few (say, 5) observations which may not be able to provide reliable estimates for the misclassification probabilities. A possible solution in this case is to use *cross validation*. A small number (say, 5 to 10) of observations are chosen at random for the test set, and the remaining are used for model estimation. This gives an unbiased but extremely variable estimate for the classification errors. In order to reduce this variance, the operation is repeated many times (say, 1000 times) and the final error estimates are obtained by averaging.

11.2   Robust discriminant analysis

It is apparent that estimators used for the simple supervised classification problem considered in the previous section may be heavily susceptible to contamination. As noted in Hawkins and McLachlan (1997), in linear discriminant analysis even a single outlier may bias the estimates of μ_c and Σ, and lead to arbitrarily poor classification rules. It can be noted that also the heteroscedastic case is prone to bias due to contamination. One of the first references about the need of robustness in discriminant analysis is Lachenbruch *et al.* (1973).

In the heteroscedastic case a simple approach to robust discriminant analysis and sample reduction amounts to use robust location and scatter estimates, stratified by group. More precisely, one can use any robust estimator for location and scatter in place of (11.1) and (11.2) to obtain robust estimates and consequently robust classification rules. Replacement of classical with robust estimates, keeping everything else fixed, is known in the literature as the "plug-in" method. As noted in Todorov and Pires (2007), many high breakdown methods can be obtained simply by using different estimators within each group.

For the homoscedastic case, a similarly simple approach is proposed in Chork and Rousseeuw (1992): one can pool together the stratified high-breakdown estimates obtained using a robust version of (11.2). This is simple and effective, but corresponds to a situation in which a breakdown point is used within each group. A fixed amount of observations are trimmed or down-weighted within each group. This is usually referred as *partial* trimming in case of zero weights to possible outliers.

Hubert and Van Driessen (2004) also use a similar plug-in approach, where μ_c and Σ_c are estimated by means of the MCD or RMCD estimator. He and Fung (2000) instead work with S estimators based on the difference between each observation and its centroid. More precisely, it is suggested to compute robust centroid estimates $\hat{\mu}_c$, $c = 1, \ldots, k$ and then use a robust S estimate for the covariance of the vector

$$\left(\sum_{c=1}^{k} u_{1c}(x_1 - \hat{\mu}_c), \sum_{c=1}^{k} u_{2c}(x_2 - \hat{\mu}_c), \ldots, \sum_{c=1}^{k} u_{nc}(x_n - \hat{\mu}_c) \right).$$

It can be seen that this directly provides an estimate of the common covariance matrix, and trimming/downweighting is impartial. Impartial trimming usually brings about advantages over partial trimming, as some groups may be more heavily contaminated than other ones.

Hawkins and McLachlan (1997) adopt a mean-shift outlier model and a trimming approach. The mean-shift model assumes that the contaminated observations arise from Gaussian distributions with the same covariance matrix of the clean observations, and an observation-specific mean. The trimming level ε is fixed, and based on the likelihood it is possible to identify the most likely contaminated observations and estimate the parameters, where $\hat{\mu}_c$, and $\hat{\Sigma}$ have very similar expressions to those obtained for `tclust`:

$$\hat{\mu}_c = \frac{\sum_{i=1}^{n} z_i u_{ic} x_i}{\sum_{i=1}^{n} z_i u_{ic}};$$

and

$$\hat{\Sigma} = c(p,\varepsilon)\frac{\sum_{c=1}^{k}\sum_{i=1}^{n} z_i u_{ic}(x_i - \hat{\mu}_c)(x_i - \hat{\mu}_c)'}{\sum_{i=1}^{n} z_i}.$$

where the indicator z_i is estimated (via maximum likelihood) and u_{ic} is known. The constant $c(p,\varepsilon)$ is a consistency factor obtained in Butler *et al.* (1993) to ensure consistency after trimming, and its expression is

$$c(p,\varepsilon) = \frac{\Pr(\chi_{p+2}^2 < \chi_{1-\varepsilon,p}^2)}{1-\varepsilon}, \tag{11.8}$$

where χ_{p+2}^2 denotes a random variable distributed like a chi-square with $p+2$ degrees of freedom, and $\chi_{1-\varepsilon,p}^2$ the $1-\varepsilon$ quantile of a χ_p^2. See also Cerioli *et al.* (2013a) on this.

Pires and Branco (2010) propose yet another method, based on projection pursuit. This method is more reliable when the covariance matrix is close to singular, as with any projection pursuit approach. These cases include situations of high dimensionality.

Many of the methods referenced above are implemented in R library `rrcov`. As far as linear discriminant analysis is concerned, the method of Pires and Branco (2010) is implemented in function `LdaPP`, while a plug-in method is in function `Linda`. In function `Linda` it is possible to choose between different uses of MCD estimates, or the Hawkins and McLachlan (1997) method. Plug-in methods to obtain quadratic discriminant analysis, based on different robust estimators of location and scatter, are instead available in function `QdaCov`.

Properties of robust discriminant analysis methods are formally studied in Croux and Joossens (2004, 2005) and Croux *et al.* (2008). Croux *et al.* (2008) show that even when there is no contamination, the loss in efficiency of many robust discriminant analysis methods is limited. The class of S estimators is also considered in Croux and Dehon (2001), where the influence function of the resulting procedure is derived. Croux and Dehon (2001) also derive the optimal estimator within the class of S estimators by minimizing the influence an outlier can have on the classification error.

A final remark is that robust discriminant analysis methods may be more suited with large sample sizes. Many robust estimators are efficient only if the sample size is at least twice the number of variables. Given that in many cases robust estimators are used *within* groups, we recommend that the size of the smallest group is twice the number of variables. This is not what will happen for the two examples that follow, anyway.

11.1 Example (Discriminant analysis of Iris data) *We begin with the famous Iris data, collected by Anderson (1935) and used by Fisher (1936) to illustrate his version of discriminant analysis. The data set gives the measurements in centimeters of $p = 4$ variables, measuring sepal length and width, and petal length and width. The measurements are made on 50 flowers from each*

TABLE 11.1
Iris data. Stratified means (upper panel) and RMCD location estimates (lower panel). L. is a short notation for "length," W. for "width."

Species	Sepal L.	Sepal W.	Petal L.	Petal W.
I. setosa	5.01	3.43	1.46	0.25
I. versicolor	5.94	2.77	4.26	1.33
I. virginica	6.59	2.97	5.55	2.03
I. setosa	4.97	3.38	1.44	0.20
I. versicolor	5.92	2.81	4.23	1.31
I. virginica	6.49	2.95	5.45	2.02

of $k = 3$ species of Iris, observed in the Gaspe peninsula, for a total of $n = 150$ observations. The species are *Iris setosa*, *I. versicolor*, and *I. virginica*.

We begin with data reduction. This is completely straightforward as the cluster labels are known in advance. We report in Table 11.1 the location estimate of each of the $k = 3$ groups. In the first panel we report the classical estimates, and in the second panel robust estimates obtained using the RMCD.

It can be seen that there is no particular evidence of contamination for these data, therefore we may expect robust and classical methods to perform similarly (with slight more uncertainty connected with robust methods due to loss of efficiency).

It can also be seen that the species *I. setosa* has a much smaller petal, on average, with respect to the other two. This allows us to distinguish and correctly classify this species. On the other hand, *I. virginica* flowers seem to be the largest in general, but there may be some overlapping with *I. versicolor*. This is confirmed by the discriminant analysis below.

An issue now is to select the best classification rule for building future predictions. To this end we implement cross validation, selecting a random sample of 120 flowers, 40 from each group, for training the classifiers. The remaining 30 are used for estimation of the misclassification error. We estimate the misclassification error as the proportion of errors among the 30 predicted labels. The operation is repeated 1000 times, and we report the average proportion of misclassified flowers for each method in Table 11.2. Given the small sample size we compare only methods for *linear* discriminant analysis. LDA stands for classical linear discriminant analysis, LDA-HM for the Hawkins and McLachlan (1997) method, and LDA-MCD for the plug-in method with RMCD with pooled within covariance matrix.

On the basis of the results in Table 11.2, we conclude that the best linear discriminant analysis approach is classical LDA. We noted that there is not much evidence of contamination with these data, yet we compared LDA with robust methods. In this case, robust methods are simply seen as different classification methods, which are evaluated in order to select the best classifier.

For the final model we report the misclassification matrix (or confusion

TABLE 11.2

Iris data. Cross-validated misclassification error and different linear discriminant analysis approaches. The results are averaged over 1000 replicates, with a test set of 30 flowers.

Method	Misclassification
LDA	0.021
LDA-HM	0.030
LDA-MCD	0.031

TABLE 11.3

Iris data. Misclassification matrix of classical LDA estimated with cross-validation. The results are based on 1000 replicates and estimate the conditional probability of classification.

	I. setosa	I. versicolor	I. virginica
I. setosa	1.00	0.00	0.00
I. versicolor	0.00	0.96	0.04
I. virginica	0.00	0.02	0.98

matrix) of LDA, once again estimated using cross-validation, in Table 11.3. In the rows we have the true and in the columns the predicted labels.

It can be seen as expected that misclassification occurs only for the *I. versicolor* and *I. virginica* species.

We conclude reporting a plot of the first two principal components of the data, which helps us in making further considerations. The biplot is in Figure 11.1, where we distinguish the three groups using different symbols.

From the biplot it is apparent that the data set only contains two, rather than three, clusters. It is indeed possible that two of the clusters overlap due to projection (while they might be separated in the four dimensional space of the original data). A natural cluster in the (best) two dimensional projection is made by both *I. virginica* and *I. versicolor*, and does not seem to be separable.

11.2 Example (Discriminant analysis of spam data) We now revisit the spam data set, where we have two groups. One is made of 30 spam e-mails, and another of 2456 non spam e-mails. A total of $p = 20$ variables are measured for each e-mail. Building a robust and effective classifier is very important with these data, as even if we have an intuition that there is an inherent difference between spam and clean e-mails, we would like to have an automatic classification method which can flag spam e-mails without our direct intervention.

First of all, it can be seen that groups are well separated with respect to the centroid. We can therefore expect classification to be generally successful. The reduced data is in Table 11.4.

We now compare classifiers as before using cross-validation. The test set

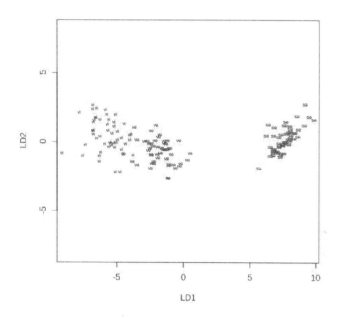

FIGURE 11.1
Iris Data. Biplot for linear discriminant analysis.

is repeatedly obtained by sampling 30% of the clean e-mails and five e-mails from the spam group. Given that the data set is larger than the previous, we also compare quadratic discriminant analysis methods.

From Table 11.5 we can see that the best method is robust quadratic discriminant analysis based on the RMCD, which obtains the smallest classification error. Given the target of this application, anyway, this does not mean that QDA-MCD is the best choice. A problem with this data set is that it is highly imbalanced, as there is little information in the data about spam. Further, false positives and false negatives have a different practical impact with this application. In this light, we compare the confusion matrices for different methods. For reasons of space we compare only LDA and QDA-MCD in Table 11.6.

It can be clearly seen that the better overall misclassification rate is achieved by QDA-MCD through classification of almost all e-mails as clean. This leads to 4% increase in the conditional probability of correctly classifying a clean e-mail with respect to classical LDA, at the price of more than 50% decrease in the probability of correctly classifying spam. Given that spam e-mails are so rare in this data set, the number of correctly classified e-mails

TABLE 11.4
Spam data. Stratified means (left panel) and RMCD location (right panel) estimates.

	Mean clean e-mails	spam	RMCD clean e-mails	spam
V1	-2.20	-0.08	-2.26	0.18
V2	-3.12	-2.22	-3.10	-2.16
V3	-3.23	-2.10	-3.18	-2.08
V4	-3.24	-2.16	-3.18	-2.10
V5	-1.17	0.19	-1.25	0.10
V6	-3.13	-1.81	-3.10	-1.70
V7	-3.08	-2.48	-3.03	-2.61
V8	-1.91	-3.30	-1.95	-3.27
V9	3.99	5.41	3.95	5.56
V10	-2.73	-1.58	-2.71	-1.56
V11	-2.88	-1.26	-2.88	-1.33
V12	-3.09	-2.40	-3.12	-2.49
V13	-2.35	-3.28	-2.34	-3.26
V14	-3.07	-2.59	-3.03	-2.56
V15	-3.15	-2.77	-3.10	-2.86
V16	-3.23	-2.18	-3.16	-2.33
V17	-3.14	-2.92	-3.11	-3.06
V18	-3.03	-2.41	-3.02	-2.47
V19	-2.59	-1.55	-2.61	-1.55
V20	-3.26	-3.26	-3.21	-3.05

overall is much larger, hence the results in Table 11.5. On the basis of results in Table 11.6 we suggest once again choosing classical LDA for future prediction.

Note that we could improve the performance of classifiers at least in two ways: first, we could penalize false negatives by working with prior probabilities. Even in this case anyway the maximum conditional probability of correct classification of spam that can be obtained is about 25%. Secondly, we have used so far a uniform threshold for classification (that is, we have assigned an observation to the closest group), but we may fine tune the threshold to target a pre-specified false positive or false negative rate. This could lead to select a value *const* so that to assign an observation to the spam group only if it is *const*-fold closer to it than to the clean group.

11.3 R Illustration (Discriminant analysis of spam data) *For the Spam example, data are stored in a data frame* sp. *Cross-validation was performed using a* for *cycle and using function* sample *at each iteration to select the training set. Two functions (*f *and* frob*) were coded to evaluate the performance of the classical and robust classifiers on the test set.*

```
> library(rrcov)
```

TABLE 11.5

Spam data. Cross-validated misclassification error for different linear discriminant analysis approaches. The results are averaged over 1000 replicates, with a test set of 30 e-mails.

Method	Misclassification
LDA	0.009
QDA	0.008
LDA-HM	0.010
LDA-MCD	0.010
QDA-MCD	0.007

TABLE 11.6

Spam data. Misclassification matrix of classical LDA and robust QDA estimated with cross-validation. The results are based on 1000 replicates and estimate the conditional probability of classification.

	LDA	
	clean	*spam*
clean	0.995	0.005
spam	0.440	0.560

	QDA-MCD	
	clean	*spam*
clean	0.990	0.001
spam	0.978	0.022

```
> library(rrcovHD)
> library(MASS)
>
> n=nrow(sp)
> out=c(rep(1,30),rep(0,n-30))
>
> B=1000
>
> f=function(obj,test) {
+ nrow(test)-sum(diag(table(predict(obj,
+ test[,-21])$class,test[,21]))))}
> frob=function(obj,test) {
+ nrow(test)-sum(diag(table(attributes(
+ unclass(predict(obj,test[,-21])))$class,
+ test[,21]))))}
```

We estimate the outlier detection performance through cross-validation as follows:

```
> errors=matrix(NA,B,5)
```

```
> roberrors=matrix(NA,B,5)
>
> for(i in 1:B) {
+ w=c(sample(30,25),30+sample(n-30,(n-30)*0.7))
+ ld=lda(out[w]~.,data=sp[w,])
+ qd=qda(out[w]~.,data=sp[w,])
+ lpi=Linda(sp[w,],out[w])
+ lpf=Linda(sp[w,],out[w],control="fsa")
+ qdr=QdaCov(sp[w,],out[w])
+ ## outlier detection in the test set ##
+ testi=sp[-w,]
+ flag=c(1:5,5+which(getFlag(
+ OutlierMahdist(testi[-c(1:5),]))==1))
+ ## errors ##
+ testi=cbind(testi,out[-w])
+ errors[i,]=c(f(ld,testi),f(qd,testi),frob(lpi,testi),
+ frob(lpf,testi),frob(qdr,testi))/nrow(testi)
+ roberrors[i,]=c(f(ld,testi[flag,]),f(qd,testi[flag,]),
+ frob(lpi,testi[flag,]),frob(lpf,testi[flag,]),
+ frob(qdr,testi[flag,]))/length(flag)}
```

We then perform outlier detection on the whole data.

```
> lp=lda(out~.,data=sp)
> qd=qda(out~.,data=sp)
>
> lpi=Linda(sp,out)
> lpf=Linda(sp,out,control="fsa")
> qdr=QdaCov(sp,out)
```

Finally, we honestly estimate the confusion matrix through a separate cross-validation study:

```
> ta=matrix(0,2,2)
> ta2=matrix(0,2,2)
>
> for(i in 1:B) {
+ w=c(sample(30,25),30+sample(n-30,(n-30)*0.7))
+ ld=lda(out[w]~.,data=sp[w,])
+ qdr=QdaCov(sp[w,],out[w])
+ ta=table(out[-w],predict(ld,sp[-w,])$class)+ta
+ ta2=table(out[-w],attributes(unclass(predict(
+ qdr,sp[-w,])))$class)+ta2}
>
> ta/apply(ta,1,sum)
> ta2/apply(ta2,1,sum)
```

A

Use of the Software R for Data Reduction

The purpose of this appendix is to give more details on how robust data reduction procedures can be implemented within the software R. We will point the reader to the relevant libraries or to source files accompanying the book where something is missing from the currently available libraries. We will also discuss the main options for fine tuning of the procedures. Of course, the interested reader is recommended to read the online manual of the functions below, when they are included in a library, with a call along the lines of `help(function)`.

A.1 Multivariate estimation methods

A large number of multivariate estimation methods are supplied in the R library `rrcov`.

S and MM estimators

S and MM estimates can be computed with the functions

```
> CovSest(x=X, bdp = 0.5, maxiter = 120)
> CovMMest(x=X, bdp = 0.5, eff = 0.95,
+ eff.shape=FALSE,maxiter=50)
```

respectively.

Data are stored in a matrix or data frame X. The user is allowed to fix the breakdown point of the initial S estimate $\varepsilon = 1 - $ bdp, and also the efficiency eff for the MM estimator.

The option `eff.shape` is a logical value indicating whether efficiency has to be determined with respect to shape or location (defaulting to FALSE and hence to location efficiency. The option `maxiter` indicates the maximum number of iterations the algorithm when computing the S estimate.

The function `CovMMest` returns both the initial S estimate and the final MM estimate.

Minimum Covariance Determinant

The raw MCD and RMCD estimates can be obtained with the function

```
> CovMcd(x=X, alpha = 0.5, nsamp = 500,
+ use.correction = TRUE)
```

The option `alpha` controls the size $h = \lfloor \alpha n \rfloor$ of the subsets over which the determinant is minimized, with $0 < \alpha \leq 0.5$. It therefore corresponds to ε in the notation of this book.

The `nsamp` argument gives the number of $(p+1)$-dimensional subsets from which the initial subset of size h is obtained. A finite sample correction is used by default for consistency.

The function returns both the raw and the reweighted MCD. The reweighting step is carried out through the χ_p^2 approximation. The raw MCD can be used to compute the RMCD based on the scaled F distribution using `CovMcdF`, that is available as online supplementary material for this book.

2GSE

The 2GSE has been introduced in Agostinelli *et al.* (2014) as an estimator resistant to component-wise outliers. It is implemented in `library(GSE)` as function `TSGS`. It has the advantage of needing basically no tuning or initialization. A simple call

```
> TSGS(X)
```

computes the robust estimates, which can be accessed through

```
> TSGS(X)@mu
> TSGS(X)@S
```

or using functions `getLocation` and `getScatter`.

snipEM

`snipEM` corresponds to `sclust` when $k = 1$. It is *optimal* in presence of cell-wise outliers, so that if the snipping level is carefully chosen and the snipping matrix `V` is initialized well it provides the best performance. The estimator can be found in library `snipEM`, as

```
> snipEM(X, V, tol = 1e-04, maxiters = 500,
+ maxiters.S = 1000, print.it = FALSE)
```

where `V` is an initial guess for location of component-wise outliers and `sum(1-V)` is the fixed snipping level $np\varepsilon$. The option `maxiters` controls the maximum number of iterations of the EM algorithm, while `maxiters.S` controls the maximum number of iterations of the greedy algorithm which is used to update V. Results are better and better as `maxiters.S` is increased.

The function returns a list with estimated location (`mu`), estimated scatter (`S`), final V matrix for identification of the snipped entries, and likelihood at convergence (`lik`).

---

stEM

A generalization of `snipEM` corresponds to simultaneously perform snipping and trimming. The resulting function is *optimal* in presence of both cell-wise and component-wise outliers, so that if the snipping and trimming levels are carefully set and the V matrix is initialized well it provides the best possible performance. See also Farcomeni (2015). The estimator can be found in library `snipEM`, as

```
> stEM(X, V, tol = 1e-04, maxiters = 500,
+ maxiters.S = 1000, print.it = FALSE)
```

where the only difference with function `snipEM` is that V must contain also rows of zeros. The number of rows of zeros fixes the trimming level and number of remaining isolated zeros fixes the snipping level.

Outlier detection

Multivariate outlier detection relies on the computation of robust distances. This task can be accomplished as

```
> rob.dist<-mahalanobis(X, center, cov)
> which(rob.dist>quant)
```

where `quant` is a pre-specified threshold for the robust distances.

The function `mahalanobis` computes Mahalanobis distances of the rows of X, with location (`center`) covariance (`cov`). The function returns the squared Mahalanobis distances when the sample mean and sample covariance matrix are plugged in, otherwise it produces robust distances corresponding to robust estimates.

The software R also provides function `identify`, which allows to highlight data points inside a graphical display. The cut-off value `quant` is chosen according to the distribution of the robust distances. According to the asymptotic χ_p^2 result that is commonly used for all robust multivariate estimators, this value is obtained as

```
> level<-0.975
> quant<-qchisq(level,p)
```

In the particular case of the RMCD described in Section 2.4, according to the outlier identification rules in (2.28) and (2.29), we should compute two cut-off values: one for the non-trimmed points (q.non.trim), the other for the trimmed data (q.trim). One can use the following code using the output from CovMcd or CovMcdF

```
> w<-sum(output@wt)
> c1<-((w-1)^2)/w
> q.non.trimm<-qbeta(level,p/2, (w-p-1)/2)*c1
> c2<-(w^2-1)*p/w/(w-p)
> q.trimm<-qf(level,p, w-p)*c2
```

Graphical tools

Robust libraries usually define plot methods. This implies that using the generic plot function on the output of functions for computing robust estimators will give useful graphical outputs. In case outlier detection is performed, all are based on the χ_p^2 approximation.

For example, for the Stars data, the MM estimate with 50% BP and 95% shape efficiency has been obtained as

```
> mm.stars<-CovMMest(starsCYG,bdp=0.5,
+ eff=0.95, eff.shape=TRUE)
```

We can display both robust and classical Mahalanobis distances in parallel panels, as shown in Figure A.1 with the simple code

```
> plot(mm.stars,which="xydistance")
```

and the corresponding quantile-quantile plots, in Figure A.2, through

```
> plot(mm.stars,which="xyqqchi2")
```

We can only plot the left hand panels in Figure A.1 and Figure A.2 by setting which=''distance'' and which=''qqchi2'', respectively; a robust tolerance ellipse is the result of specifying option which=''tolEllipsePlot''; the distance-distance plot is obtained by specifying which=''dd''. The robust and classic tolerance ellipses, displayed in Figure A.3, can be plotted together by setting which=''pairs'' and they are supplied with the corresponding correlation estimates and marginal distributions. Robust correlations can be obtained through function getCorr:

```
> getCorr(mm.stars)
```

Bivariate tolerance ellipses are obtained through the function ellipse from package ellipse. An ellipse can be add to an existing bivariate scatterplot with the lines function:

```
> lines(ellipse(cov,center,level=0.975,
+ t= sqrt(quantile)),lty=1)
```

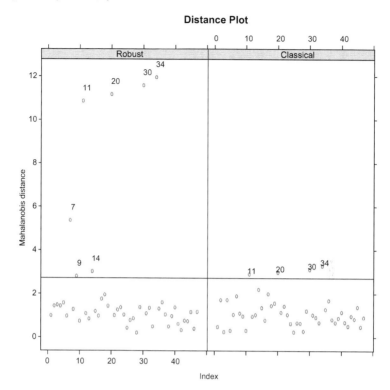

FIGURE A.1
Stars data. Robust and Mahalanobis distances.

In the code above, the user specifies a location center and covariance cov. Then, the option level controls the size of the ellipse. A vector of levels may be used to draw more than one ellipse with the same center and scatter. The option t allows to specify the appropriate quantile from a reference distribution. This is very useful as a different reference distribution rather than the χ^2_2 might be used (for instance, the scaled Beta distribution).

A.2 Robust PCA

Classical PCA can be performed in R using function princomp.

Functions to perform robust PCA based on a robust estimate of multivariate location and covariance are included in library rrcov. Projection pursuit techniques are available from package pcaPP.

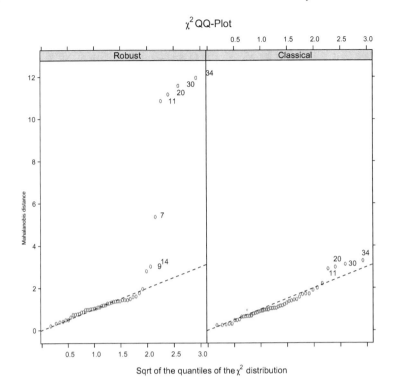

FIGURE A.2
Stars data. Quantile-quantile plot for robust and Mahalanobis distances.

PCA based on a robust covariance estimate

Robust PCA based on a robust covariance estimate can be obtained with function PcaCov, whose syntax is

```
> PcaCov(x=X, k=q,
+ cov.control=CovControlMcd(alpha=0.75),
+ scale = TRUE)
```

The value of k is the number of components to compute (q in the notation of this book). The option cov.control is used to select the robust estimate of covariance (here, we have specified the RMCD with 25% BP), scale indicates whether the variables should be scaled (MAD is used as default but, alternatively, other measures can be supplied).

The output includes robust estimates of eigenvalues, loadings, scores, location, score and orthogonal distances. The plot function can be used to draw outlier maps.

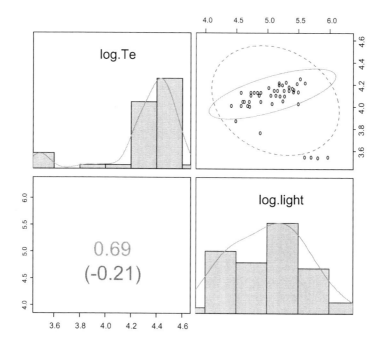

FIGURE A.3
Stars data. Marginal distributions (top left and bottom right); robust and classical tolerance ellipses (top right); robust and classical estimate of the correlation coefficient (bottom left).

Spherical PCA and ROBPCA

Spherical PCA can be obtained with the following code:

```
> PcaLocantore(x=X, k=q, scale = TRUE)
```

while ROBPCA through

```
> PcaHubert(x=X, k=q, kmax, alpha=0.75,
+ scale = FALSE)
```

where k and kmax are used to specify the maximum number of components to be computed (q in the notation of this book), one minus alpha corresponds to the BP. Finally, scale is a logical value indicating whether variables should be robustly scaled.

The output includes score and orthogonal distances, with the corresponding cut-off values based on chi-squared approximations.

PCA based on projection pursuit

The original algorithm of Croux and Ruiz-Gazen (2005) and its improvement based on the grid search algorithm (Croux *et al.*, 2007) can be found in the `pcaPP` library by calling the functions

```
> PCAproj(x, k = 2, method = c("mad", "sd", "qn"),
+ scores = TRUE)
```

and

```
> PCAgrid(x, k = 2, method = c("mad", "sd", "qn"),
+ scores = TRUE)
```

respectively. The user can choose the projection index through one of the three options for `method` (i.e., only one among `mad`, `sd` and `qn` will be specified). Additionally one can decide whether to store the scores (defaulting to `TRUE`). Additional options can be used to select ways to center and scale the data before projection pursuit.

Score and orthogonal distances, along with the corresponding cut-off values, can be obtained through the function

```
> PCdiagplot(x,ppobj, raw=FALSE)
```

where `ppobj` is the object containing the output from one of the two functions mentioned above, and `raw` is a logical value indicating whether the robust score distances have to be adjusted to improve the χ^2_k approximation. If `FALSE`, a factor is used in order to improve the approximation; otherwise the raw distances are used.

A.3 Sparse robust PCA

---

sPCA

The algorithm for sPCA can be found in function `spca` from package `elasticnet`. Robust sPCA is obtained if a robust estimate of covariance is given as input. The general syntax is

```
> spca(x=S, K=q, para=rep(5,q),type="Gram",
+ sparse='varnum')
```

where K is used to specify the number of sparse components to compute
(*q* in the notation of this book). Option type=''Gram'' is used when x
is a covariance matrix. When the data matrix is supplied, we need to set
type=''predict'' (and in this case sPCA is not robust).

The option para allows to fix the values of the penalty parameters *or* the
number of nonnull loadings for each component. It is a vector of length K. In
sparse = ''penalty'', then para gives a vector of penalty parameters for
the L_1 penalty. If sparse = ''varnum'', then (like in the example above)
the number of nonnull loadings is specified for each of the K components.

The output includes the sparse loadings along with the percentage of explained variance.

Sparse projection pursuit

Sparse and robust PCA procedures based on projection pursuit are supplied in library pcaPP. They are obtained with the following code:

```
sPCAgrid(X, k=q,method="mad",lambda)
```

where as usual k is the number of components, and method is used to choose
the projection index (possible values are sd, mad and qn). The parameter
lambda can be a scalar or a vector of length k, giving the penalty value in-
ducing sparseness. In case a scalar is specified, the same penalty is used for
all components. The penalty can be chosen using function opt.TPO.

A.4   Canonical correlation analysis

Classical Canonical Correlation Analysis can be performed in R through the
function cancor. Supplemental outputs and graphical devices can be obtained
from function cc in library CCA.

Robust CCA based on projection pursuit can be obtained through function
ccaProj in library ccaPP, with syntax

```
> ccaProj(x, y, k = q, method = "pearson")
```

where x and y are vectors, matrices or data frames; k is the number of canonical variables to compute and method is the correlation functional to use.
Possible values include pearson for classical Pearson correlation, kendall
and spearman for these two correlation coefficients, quadrant for quadrant
correlation and M for a bivariate M estimator of correlation based on Huber
function.

Other robust counterparts of classical CCA shall at the moment be obtained manually by eigen analysis of the appropriate product matrices after
computing robust estimates of covariance.

A.5   Factor analysis

Classical Factor Analysis based on maximum likelihood is supplied by function
`factanal`. Other estimation methods are available through function `fa` in
package `psych`. The syntax of the latter is

```
> fa(r=R, nfactors=q, rotate="none")
```

In the syntax above, `r` is a covariance or correlation matrix, `nfactors` denotes
the number of common factors to compute. Option `rotate` is used select the
type of rotation, where `none` indicates no rotation (which is *not* the default).
Among the possible options, we mention `varimax` and `oblimin`.

The `fa` function can be directly used for robust FA, simply by supplying a
robust estimate of covariance or correlation. Robust estimates of correlation
are obtained either by robust standardization before estimation, or through
function `cov2cor`.

Library `robusfa` provides tools for robust FA. It relies on packages `rrcov`,
`robustbase` and `pcaPP`. A robust factor model can be fit to the data x using
the function `FaCov`.

```
> FaCov(x, k=q,
+ cov.control=CovControlMcd(alpha=0.75),
+ scale = TRUE)
```

where `cov.control` allows to choose the robust estimate of covariance (the
RMCD is the default option), and `scale` is a logical value indicating whether
the data have to be scaled by their column-wise MAD.

A.6   Classical k-means and model based clustering

For clustering, k-means is implemented in function `kmeans` available from the
R base. Library `cluster` provides many different clustering methods mostly
taken from Kaufman and Rousseeuw (1990), including `pam`. Library `mclust`
can be used to compute model based clustering methods. We now give few
details on these functions.

k-means

The function `kmeans` for data x and k groups can be called with the following syntax:

```
> kmeans(x, k, iter.max=100, n.start=20)
```

where `iter.max` allows the algorithm to run for a maximum of 100 iterations, and `n.start` makes the function try 20 initial random starts. These parameters can be tuned to trade off computational time and likelihood of obtaining the global optimum.

Model Based Clustering

Model based clustering with unconstrained covariance matrices and k groups can be obtained with library `mclust`, calling

```
> Mclust(x, G=k, modelNames="VVV")
```

It shall be noted that a range of models can be compared based on the BIC simply by omitting the option `modelNames` or by specifying a vector of model names. Model names correspond to those given in Table 7.3.

Option `G` controls the number of clusters. If it is omitted, a range of possible values for the number of clusters is estimated, and the results are compared via the BIC.

It is important here to underline that eigenvalues of covariance matrices are not constrained using function `Mclust`, therefore it is possible to obtain spurious solutions, especially with the *VVV* model. In order to avoid spurious solutions, a simple possibility with the *VVV* models is to use `tclust` with no trimming, using library `tclust` and then

```
> tclust(x,k,alpha=0)
```

Cluster Evaluation

Library `cluster` contains the function `silhouette` which can be easily used to compute the silhouette statistics.

A number of other statistics can be found in function `cluster.stats` within library `fpv`, while the GAP statistic can be computed for instance using the function `clusGap` in library `cluster`.

We provide code for the adjusted Rand index, which can also be obtained with function `adjustedRandIndex` in library `mclust`.

A.7 Robust clustering

The PAM algorithm is found in function `pam` in library `cluster`. Library `tclust` provides trimmed k-means (function `tkmeans`) and the `tclust` algorithm, and additionally the trimmed likelihood curves in function `ctlcurves`. Snipping methods are found in library `snipEM`, which provides functions skmeans for snipped k-means and sclust for snipped robust model based clustering.

Trimmed k-means & tclust

A call to function tkmeans is along the lines of

```
> tkmeans (x, k, alpha = 0.05, nstart = 50,
+ iter.max = 20)
```

where the options specified above are actually the defaults. The parameter alpha controls the trimming proportion, while nstart the number of random initial solutions. Finally, iter.max may be tuned as usual.

The syntax for tclust is equivalent, with the additional parameter restr.fact which allows to specify a bound for the ratio between the largest and smallest eigenvalues. The default is 12.

Snipped k-means & sclust

Snipped k-means can be performed through function skmeans from library snipEM.

```
> skmeans(as.matrix(x),k,V=Z,clust=R,itersmax=10^5,
+ s=rep(1,n),D=0.1)
```

First of all, it shall be noted that the function skmeans works only with data defined as matrices (not data frames). Secondly, the number of iterations is specified using itersmax. This is usually large to guarantee convergence to the global optimum. A good initial solution is crucial in some cases (e.g., moderately large data matrices) even if itersmax is very large.

The initial solution is specified using the options V for the binary matrix of snipped entries (Z in the notation of this book), where a zero corresponds to a snipped entry, and clust for the initial cluster labels. Initial cluster labels must be integers from 1 to k, with at least one observation assigned to each group.

If there are trimmed entries in Z, these can be specified providing an appropriate binary vector s. The argument s can be missing, defaulting to rep(1,n) (no trimming). Note that we do not specify the snipping level ε as the latter is a consequence of the proportion of zeros in V. Finally, we recommend repeatedly running the algorithm with different values of the tuning parameter D (*const* in the notation of this book).

The syntax for sclust is similar, where the function can once again be found in library snipEM. This can be called using

```
> sclust(as.matrix(x),k,V,R,restr.fact=12,tol=1e-5,
+ maxiters = 100, maxiters.S = 1000)
```

where V and R are once again initial solutions for the snipped entries and clustering, respectively; `restr.fact` is the bound for the ratio of eigenvalues, `tol` is the convergence criterion for the likelihood and `maxiters` is the maximum number of iterations of the EM. An important tuning parameter is `maxiters.S`, which controls the iterations of the stochastic optimization algorithm repeated at each S-step. Larger values lead to better solutions but at the price of a larger computational time.

A.8 Robust double clustering

We make available one function to perform trimmed double k-means and another to perform snipped double k-means in the online supplementary material of this book.

Trimmed double k-means

The function to perform the trimmed double k-means procedure requires specification of the *number*, rather than proportion, of outliers. These are separately specified for rows and columns as follows:

```
> trimdkm(X,k1,k2,nr,nc)
```

where k1 and k2 denote the number of row and column clusters, respectively, and nr and nc denote the number of row and column outliers to be trimmed. The function returns a list with elements m, corresponding to the estimated centroid matrix $\bar{X}$; rowp and colp, corresponding to the estimated row and column cluster labels, respectively; and obj corresponding to the value of the objective function at convergence.

The default is to initialize the partitions using random sampling with replacement. Option `initial=2` can be specified to initialize by estimating the row partition using PAM, and similarly for the columns. It is recommended that the algorithm is repeatedly performed with random initialization to increase the likelihood of obtaining the global optimum. This is easily performed with a cycle along the lines of

```
> res = trimdkm(X,k1,k2,nr,nc)
> for(i in 1:50) {
>     res2 = trimdkm(X,k1,k2,nr,nc)
>     if(res2$obj<res$obj) {
>             res=res2}}
```

Snipped double k-means

Snipped double k-means can be performed with a code along the lines of

```
> snipdkm(X,k1,k2,Z,D=0.1,maxiter=100,tol=1e-6,
+ inits=NULL)
```

where `Z` denotes a binary matrix with zeros in correspondence of initially snipped entries. The fraction of snipped entries is in one-to-one correspondence with the proportions of zeros in `Z`, and is therefore not specified. A deterministic initialization of the row and column partitions can be specified by passing a list to `inits`, otherwise these are randomly initialized as usual. The parameters `D` and `maxiter` control the stochastic optimizer as usual, while `tol` is the tolerance for convergence in the objective function. Once again, a multistart is recommended.

A.9 Discriminant analysis

Classical discriminant analysis methods can be found in library `MASS`, while few robust methods are available in library `rrcov`.

Classical discriminant analysis

Linear discriminant analysis is obtained with

```
> lda(x,R)
```

where `x` is a matrix or data frame of predictors, and `R` contains the group labels from 1 to k. The option `CV=TRUE` can be used to perform leave-one-out cross validation. Function `predict` is used to predict new data with the usual R syntax. The outcome of `lda` can be passed to function `plot` to obtain a biplot. Function `qda` gives quadratic discriminant analysis and has the same syntax.

Robust discriminant analysis

Robust discriminant analysis as proposed by Hawkins and McLachlan (1997) is obtained as:

```
> Linda(x,R,control="fsa")
```

while option `control=''MCD''`, which is the default, gives a plug-in method based on the MCD and pooled covariance matrix.

A function `LdaPP` gives robust discriminant analysis based on projection pursuit, but is momentarily restricted to $k = 2$. Quadratic discriminant analysis based on the MCD is obtained with function `QdaCov`.

Bibliography

J. G. ADROVER (1998). Minimax bias-robust estimation of the dispersion matrix of a multivariate distribution. *Annals of statistics*, 2301–2320.

C. AGOSTINELLI AND L. GRECO (2013). A weighted strategy to handle likelihood uncertainty in Bayesian inference. *Computational Statistics*, **28**, 319–339.

C. AGOSTINELLI, A. LEUNG, V. J. YOHAI, AND R. H. ZAMAR (2014). Robust estimation of multivariate location and scatter in the presence of cellwise and casewise contamination. *TEST*, in press.

J. AGULLÓ, C. CROUX, AND S. VAN AELST (2008). The multivariate least-trimmed squares estimator. *Journal of Multivariate Analysis*, **99**, 311–338.

F. AIRES, A. CHEDIN, AND J.P. NADAL (2000). Independent component analysis of multivariate time series: Application to tropical SST variability. *Journal of Geophysical Research*, **105**, 437–455.

H. AKAIKE (1973). Information theory as an extension of the maximum likelihood principle. In: B. N. PETROV AND CSAKI F., eds., *Second International symposium on information theory*, 267–281. Akademiai Kiado, Budapest.

A. D. AKKAYA AND M. L. TIKU (2005). Robust estimation and hypothesis testing under short-tailedness and inliers. *TEST*, **14**, 129–150.

A. D. AKKAYA AND M. L. TIKU (2008). Short-tailed distributions and inliers. *TEST*, **17**, 282–296.

M. ALFÓ, A. FARCOMENI, AND L. TARDELLA (2007). Robust semiparametric mixing for detecting differentially expressed genes in microarray experiments. *Computational Statistics & Data Analysis*, **51**, 5253–5265.

M. ALFÓ, A. FARCOMENI, AND L. TARDELLA (2011). A three component latent class model for robust semiparametric gene discovery. *Statistical Applications in Genetics and Molecular Biology*, **10**, article 7.

F. ALQALLAF, S. VAN AELST, V. J. YOHAI, AND R. H. ZAMAR (2009). Propagation of outliers in multivariate data. *Annals of Statistics*, **37**, 311–331.

M. ANDERBERG (1973). *Cluster analysis for applications*. Academic Press, New York.

E. ANDERSON (1935). The irises of the Gaspe peninsula. *Bulletin of the American Iris Society*, **59**, 2–5.

D. W. APLEY AND J. SHI (2001). A factor-analysis method for diagnosing variability in multivariate manufacturing processes. *Technometrics*, **43**, 84–95.

F. ATTORRE, A. ISSA, L. MALATESTA, A. ADEEB, M. DE SANCTIS, M. VITALE, AND A. FARCOMENI (2014). Analysing the relationship between land units and plant communities: the case of Socotra island (Yemen). *Plant Biosystems*, **148**, 529–539.

K. BACHE AND M. LICHMAN (2013). UCI machine learning repository.

A. BANERJEE AND R.N. DAVE (2012). Robust clustering. *WIREs Data Mining and Knowledge Discovery*, **2**, 29–59.

J. BANFIELD AND A. RAFTERY (1993). Model-based Gaussian and non-Gaussian clustering. *Biometrics*, **49**, 803–821.

V. BARNETT AND T. LEWIS (1994). *Outliers in Statistical Data*. Wiley, New York.

M. BELKIN, P. NIYOGI, AND V. SINDHWANI (2006). Manifold regularization: a geometric framework for learning from labeled and unlabeled examples. *The Journal of Machine Learning Research*, **7**, 2399–2434.

Y. BENJAMINI AND Y. HOCHBERG (1995). Controlling the false discovery rate: A practical and powerful approach to multiple testing. *Journal of the Royal Statistical Society (Series B)*, **57**, 289–300.

C. A. BENNET (1954). Effect of measurement error on chemical process control. *Industrial Quality Control*, **11**, 17–20.

J. C. BEZDEK, R. J. HATHAWAY, R. E. HOWARD, C. A. WILSON, AND M. P. WINDHAM (1987). Local convergence analysis of a grouped variable version of coordinate descent. *Journal of Optimization Theory and Application*, **54**, 471–477.

C. BIERNACKI, G. CELEUX, AND G. GOVAERT (2000). Assessing a mixture model for clustering with the integrated completed likelihood. *IEEE Transactions on Pattern Analysis and Machine Intelligence*, **22**, 719–725.

H.-H. BOCK (1996). Probabilistic models in cluster analysis. *Computational Statistics & Data Analysis*, **23**, 5–28.

J. A. BRANCO, C. CROUX, P. FILZMOSER, AND M. R. OLIVEIRA (2005). Robust canonical correlations: A comparative study. *Computational Statistics*, **20**, 203–229.

D. BRIGO AND F. MERCURIO (2001). *Interest rate models: theory and practice*. Springer.

P. BRYANT (1991). Large-sample results for optimization-based clustering methods. *Journal of Classification*, **8**, 31–44.

K. P. BURNHAM AND D. R. ANDERSON (2002). *Model selection and multi-model inference: a practical information-theoretic approach*. Springer-Verlag, New York.

R. W. BUTLER, P. L. DAVIES, AND M. JHUN (1993). Asymptotics for the minimum covariance determinant estimator. *Annals of Statistics*, **21**, 1385–1400.

J. CADIMA AND I. T. JOLLIFFE (1995). Loading and correlations in the interpretation of principal components. *Journal of Applied Statistics*, **22**, 203–214.

N.A. CAMBPELL (1984). Mixture models and atypical values. *Mathematical Geology*, **16**, 465–477.

A. CAPALBO, F. SAGNELLA, R. APA, A. M. FULGHESU, A. LANZONE, A. MORCIANO, A. FARCOMENI, M. F. GANGALE, F. MORO, D. MARTINEZ, F. SPETTU, A. CAPPAI, C. CARCASSI, G. NERI, AND F. D. TIZIANO (2012). The 312N variant of luteinizing hormone/choriogonadotropin receptor gene (LHCGR) confers up to 2.7-fold increase of risk of polycystic ovary syndrome in a Sardinian population. *Clinical Endocrinology*, **77**, 113–119.

J. CARMICHAEL, J. A. GEORGE, AND R. JULIUS (1968). Finding natural clusters. *Systematic Zoology*, **17**, 144–150.

M. CARREIRA-PERPINAN AND C. WILLIAMS (2003). On the number of modes of a Gaussian mixture. In: *Scale Space Methods in Computer Vision*, 625–640. Springer.

E. A. CATOR AND H. P. LOPUHAÄ (2012). Central limit theorem and influence function for the MCD estimators at general multivariate distributions. *Bernoulli*, **18**, 520–551.

G. CELEUX AND G. GOVAERT (1992). A classification EM algorithm for clustering and two stochastic versions. *Computational Statistics & Data Analysis*, **14**, 315–332.

G. CELEUX AND G. GOVAERT (1993). Comparison of the mixture and the classification maximum likelihood in cluster analysis. *Journal of Statistical Computation and Simulation*, **47**, 127–146.

G. CELEUX AND G. GOVAERT (1995). Gaussian parsimonious clustering models. *Pattern Recognition*, **28**, 781–793.

A. CERIOLI (2010). Multivariate outlier detection with high-breakdown estimators. *Journal of the American Statistical Association*, **105**, 147–156.

A. CERIOLI AND A. FARCOMENI (2011). Error rates for multivariate outlier detection. *Computational Statistics & Data Analysis*, **55**, 544–553.

A. CERIOLI, A. FARCOMENI, AND M. RIANI (2013a). Robust distances for outlier free goodness-of-fit testing. *Computational Statistics & Data Analysis*, **65**, 29–45.

A. CERIOLI, A. FARCOMENI, AND M. RIANI (2014). Strong consistency and robustness of the forward search estimator of multivariate location and scatter. *Journal of Multivariate Analysis*, **126**, 167–183.

A. CERIOLI, M. RIANI, AND F. TORTI (2013b). Size and power of multivariate outlier detection rules. In: *Algorithms from and for Nature and Life*, 3–17. Springer.

S. S. CHAE, J. L. DUBIEN, AND W. D. WARDE (2006). A method of predicting the number of clusters using Rand's statistic. *Computational Statistics & Data Analysis*, **50**, 3531–3546.

B. CHAKRABORTY AND P. CHAUDHURY (2008). On an optimization problem in robust statistics. *Journal of Computational and Graphical Statistics*, **17**, 683–702.

O. CHAPELLE, B. SCHÖLKOPF, AND A. ZIEN (2006). *Semi-supervised learning*. MIT press, Cambridge, MA.

C. CHATFIELD AND A. J. COLLINS (1980). *Introduction to Multivariate Analysis*. Chapman and Hall.

N. V. CHAWLA AND G. KARAKOULAS (2005). Learning from labeled and unlabeled data: an empirical study across techniques and domains. *Journal of Artificial Intelligence Research*, **23**, 331–366.

Y. CHENG (1995). Mean shift, mode seeking, and clustering. *IEEE Transactions on Pattern Analysis and Machine Intelligence*, **17**, 790–799.

H. CHO, I.S. DHILLON, Y. GUAN, AND S. SRA (2004). Minimum sum-squared residues co-clustering of gene expression data. *Proceedings of the Fourth SIAM International Conference of Data Mining*, 114–125.

C. Y. CHORK AND P. J. ROUSSEEUW (1992). Integrating a high breakdown option into discriminant analysis in exploration geochemistry. *Journal of Geochemical Exploration*, **43**, 191–203.

M. CHRISTOPHER (1969). Cluster analysis and market segmentation. *British Journal of Marketing*, **3**, 99–102.

J. CLATWORTHY, D. BUICK, M. HANKINS, J. WEINMAN, AND R. HORNE (2005). The use and reporting of cluster analysis in health psychology: a review. *British Journal of Health Psychology*, **10**, 329–358.

S. CLIMER AND W. ZHANG (2006). Rearrangement clustering: pitfalls, remedies, and applications. *Journal of Machine Learning Research*, **7**, 919–943.

J. COLLINS, D. JAUFER, P. VLACHOS, B. BUTLER, AND I. SUGURU (2004). Detecting collaborations in text comparing the authors' rhetorical language choices in The Federalist Papers. *Computers and the Humanities*, **38**, 15–36.

D. COMANICIU AND P. MEER (2002). Mean shift: a robust approach toward feature space analysis. *IEEE Transactions on Pattern Analysis and Machine Intelligence*, **24**, 603–619.

L. F. CRANOR AND B. A. LA MACCHIA (1998). Spam! *Communications of the ACM*, **41**, 74–83.

C. CROUX AND C. DEHON (2001). Robust linear discriminant analysis using S-estimators. *Canadian Journal of Statistics*, **29**, 473–492.

C. CROUX AND P. EXTERKATE (2011). Sparse and robust factor modeling. *Tinbergen Institute Discussion Paper TI 122/4*.

C. CROUX, P. FILZMOSER, AND H. FRITZ (2013). Robust sparse principal component analysis. *Technometrics*, **55**, 202–214.

C. CROUX, P. FILZMOSER, AND K. JOOSSENS (2008). Classification efficiencies for robust discriminant analysis. *Statistica Sinica*, **18**, 581–599.

C. CROUX, P. FILZMOSER, AND M. R. OLIVEIRA (2007). Algorithms for projection–pursuit robust principal component analysis. *Chemometrics and Intelligent Laboratory Systems*, **87**, 218–225.

C. CROUX, P. FILZMOSER, G. PISON, AND P. J. ROUSSEEUW (2003). Fitting multiplicative models by robust alternating regressions. *Statistics and Computing*, **13**, 23–36.

C. CROUX AND G. HAESBROECK (1999). Influence function and efficiency of the minimum covariance determinant scatter matrix estimator. *Journal of Multivariate Analysis*, **71**, 161–190.

C. Croux and G. Haesbroeck (2000). Principal component analysis based on robust estimators of the covariance or correlation matrix: influence functions and efficiencies. *Biometrika*, **87**, 603–618.

C. Croux and G. Haesbroeck (2001). Maxbias curves of robust scale estimators based on subranges. *Metrika*, **53**, 101–122.

C. Croux and G. Haesbroeck (2002). Maxbias curves of robust location estimators based on subranges. *Journal of Nonparametric Statistics*, **14**, 295–306.

C. Croux and K. Joossens (2004). Empirical comparison of the classification performance of robust linear and quadratic discriminant analysis. In: M. Hubert, G. Pison, A. Struyf, and S. van Aelst, eds., *Theory and Applications of Recent Robust Methods*, 131–140. Birkhauser, Basel.

C. Croux and K. Joossens (2005). Influence of observations on the misclassification probability in quadratic discriminant analysis. *Journal of Multivariate Analysis*, **96**, 384–403.

C. Croux and P. J. Rousseeuw (1992a). A class of high-breakdown scale estimators based on subranges. *Communications in statistics-theory and methods*, **21**, 1935–1951.

C. Croux and P. J. Rousseeuw (1992b). *Time-efficient algorithms for two highly robust estimators of scale*. Springer.

C. Croux and A. Ruiz-Gazen (2005). High breakdown estimators for principal components: the projection-pursuit approach revisited. *Journal of Multivariate Analysis*, **95**, 206–226.

J. A. Cuesta-Albertos, A. Gordaliza, and C. Matrán (1997). Trimmed k-means: an attempt to robustify quantizers. *Annals of Statistics*, **25**, 553–576.

J. A. Cuesta-Albertos, C. Matran, and A. Mayo-Iscar (2008). Robust estimation in the normal mixture model based on robust clustering. *Journal of the Royal Statistical Society (Series B)*, **70**, 779–802.

E. E. Cureton and R. B. D'Agostino (2013). *Factor analysis: An applied approach*. Psychology Press.

A. d'Aspremont, F. Bach, and L.E. Ghaoui (2008). Optimal solutions for sparse principal component analysis. *The Journal of Machine Learning Research*, **9**, 1269–1294.

A. d'Aspremont, F. Bach, and L.E. Ghaoui (2014). Approximation bounds for sparse principal component analysis. *Mathematical Programming*, 1–22.

P. L. DAVIES (1987). Asymptotic behaviour of S-estimates of multivariate location parameters and dispersion matrices. *Annals of Statistics*, **15**, 1269–1292.

P. L. DAVIES (1992). The asymptotics of Rousseeuw's minimum volume ellipsoid estimator. *The Annals of Statistics*, **20**, 1828–1843.

P. L. DAVIES AND U. GATHER (1993). The identification of multiple outliers. *Journal of the American Statistical Association*, **88**, 782–792.

F. DE LA TORRE AND M. J. BLACK (2003). A framework for robust subspace learning. *International Journal of Computer Vision*, **54**, 117–142.

M. DE SANCTIS, A. ADEEB, A. FARCOMENI, C. PATRIARCA, A. SAED, AND F. ATTORRE (2013). Classification and distribution patterns of plant communities on Socotra island, Yemen. *Applied Vegetation Science*, **16**, 148–165.

M. DEBRUYNE AND M. HUBERT (2009). The influence function of the Stahel-Donoho covariance estimator of smallest outlyingness. *Statistics and Probability Letters*, **79**, 275–282.

C. DEHON, P. FILZMOSER, AND C. CROUX (2000). Robust methods for canonical correlation analysis. In: *Data analysis, classification, and related methods*, 321–326. Springer.

A. P. DEMPSTER, N. M. LAIRD, AND D. B. RUBIN (1977). Maximum likelihood from incomplete data via the EM algorithm (with discussion). *Journal of the Royal Statistical Society (Series B)*, **39**, 1–38.

P. D'HAESELEER (2005). How does gene expression clustering work? *Nature Biotechnology*, **23**, 1499 – 1501.

W. R. DILLON AND M. GOLDSTEIN (1984). *Multivariate analysis: Methods and applications*. John Wiley & Sons, New York.

D. L. DONOHO (1982). Breakdown properties of multivariate location estimators. *Tech. rep.*, Harvard University, Boston.

D. L. DONOHO AND P. J. HUBER (1983). The notion of breakdown point. In: P. BICKEL, K. DOKSUM, AND J. L. JR. HODGES, eds., *A Festschrift for Erich L. Lehmann*, 157–184. Wadsworth.

R. DYKSTRA (1983). An algorithm for restricted least squares regression. *Journal of the American Statistical Association*, **78**, 837–842.

M. EISEN, P. SPELLMAN, P. BROWN, AND D. BOTSTEIN (1998). Cluster analysis and display of genome-wide expression patterns. *Proceedings of the National Academy of Sciences of the United States of America*, **95**, 14863–14868.

S. Engelen, M. Hubert, and K.V. Branden (2005). A comparison of three procedures for robust PCA in high dimensions. *Austrian Journal of Statistics*, **34**, 117–126.

B. S. Everitt, S. Landau, M. Leese, and D. Stahl (2011). *Cluster Analysis*. Wiley, Chichester.

A. Farcomeni (2008). A review of modern multiple hypothesis testing, with particular attention to the false discovery proportion. *Statistical Methods in Medical Research*, **17**, 347–388.

A. Farcomeni (2009a). An exact approach to sparse principal component analysis. *Computational Statistics*, **24**, 583–604.

A. Farcomeni (2009b). Robust double clustering: a method based on alternating concentration steps. *Journal of Classification*, **26**, 77–101.

A. Farcomeni (2014a). Robust constrained clustering in presence of entry-wise outliers. *Technometrics*, **56**, 102–111.

A. Farcomeni (2014b). Snipping for robust k-means clustering under component-wise contamination. *Statistics and Computing*, **24**, 909–917.

A. Farcomeni (2015). Comments on: Robust estimation of multivariate location and scatter in the presence of cellwise and casewise contamination. *TEST*, in press.

A. Farcomeni and S. Arima (2012). A Bayesian autoregressive three-state hidden Markov model for identifying switching monotonic regimes in microarray time course data. *Statistical Applications in Genetics and Molecular Biology*, **11**, article 3.

A. Farcomeni and L. Greco (2014). S-estimation of hidden Markov models. *Computational Statistics*, in press.

A. Farcomeni, S. Serranti, and G. Bonifazi (2008). Nonparametric analysis of infrared spectra for recognition of glass and ceramic glass fragments in recycling plants. *Waste Management*, **28**, 557–564.

A. Farcomeni and L. Ventura (2012). An overview of robust methods in medical research. *Statistical Methods in Medical Research*, **21**, 111–133.

A. Farcomeni and S. Viviani (2011). Robust estimation for the Cox regression model based on trimming. *Biometrical Journal*, **53**, 956–973.

W. H. Fellner (1986). Robust estimation of variance components. *Technometrics*, **28**, 51–60.

P. Filzmoser (1999). Robust principal component and factor analysis in the geostatistical treatment of environmental data. *Environmetrics*, **10**, 363–375.

P. FILZMOSER, R. MARONNA, AND M. WERNER (2008). Outlier identification in high dimensions. *Computational Statistics & Data Analysis*, **52**, 1694–1711.

R. A. FISHER (1935). *The design of experiments*. Oliver & Boyd.

R. A. FISHER (1936). The use of multiple measurements in taxonomic problems. *Annals of Eugenics*, **7**, 179–188.

W. FISHER (1969). *Clustering and aggregation in economics*. Johns Hopkins.

P. A. FORERO, V. KEKATOS, AND G. B. GIANNAKIS (2012). Robust clustering using outlier-sparsity regularization. *IEEE Transactions on Signal Processing*, **60**, 4163–4177.

C. FRALEY AND A. E. RAFTERY (1998). How many clusters? which clustering method? - answers via model-based cluster analysis. *Journal of the American Statistical Association*, **41**, 578–588.

C. FRALEY AND A. E. RAFTERY (1999). MCLUST: software for model-based cluster analysis. *Journal of Classification*, **16**, 297–306.

C. FRALEY AND A. E. RAFTERY (2002). Model based clustering, discriminant analysis, and density estimation. *Journal of the American Statistical Association*, **97**, 611–631.

C. FRALEY AND A. E. RAFTERY (2003). Enhanced software for model-based clustering, density estimation, and discriminant analysis: MCLUST. *Journal of Classification*, **20**, 263–286.

A. FRANK AND A. ASUNCION (2010). UCI machine learning repository.

R. FRANK AND P. GREEN (1968). Numerical taxonomy in marketing analysis: a review article. *Journal of Marketing Research*, **5**, 83–94.

H. FRITZ, L. A. GARCÍA-ESCUDERO, AND A. MAYO-ISCAR (2012). tclust: an R package for a trimming approach to cluster analysis. *Journal of Statistical Software*, **47**, 1–26.

H. FRITZ, L. A. GARCÍA-ESCUDERO, AND A. MAYO-ISCAR (2013). A fast algorithm for robust constrained clustering. *Computational Statistics & Data Analysis*, **61**, 124–136.

S. FRÜHWIRTH-SCHNATTER AND S. PYNE (2010). Bayesian inference for finite mixtures of univariate skew-normal and skew-t distributions. *Biostatistics*, **11**, 317–336.

M. T. GALLEGOS AND G. RITTER (2005). A robust method for cluster analysis. *Annals of Statistics*, **33**, 347–380.

M. T. GALLEGOS AND G. RITTER (2009a). Trimmed ML estimation of contaminated mixtures. *Sankhyā*, **71**, 164–220.

M. T. GALLEGOS AND G. RITTER (2009b). Trimming algorithms for clustering contaminated grouped data and their robustness. *Advances in Data Analysis and Classification*, **3**, 135–167.

M. T. GALLEGOS AND G. RITTER (2010). Using combinatorial optimization in model-based trimmed clustering with cardinality constraints. *Computational Statistics & Data Analyisis*, **54**, 637–654.

G. GAN, C. MA, AND J. WU (2007). *Data clustering: theory, algorithms, and applications*. Asa-Siam Series on Statistics and Applied Probability, Society for Industrial & Applied Mathematics, USA.

S. GANESALINGAM AND G. J. MCLACHLAN (1978). The efficiency of a linear discriminant function based on unclassified initial samples. *Biometrika*, **65**, 658–662.

L. A. GARCÍA-ESCUDERO AND A. GORDALIZA (1999). Robustness properties of k means and trimmed k means. *Journal of the American Statistical Association*, **94**, 956–969.

L. A. GARCÍA-ESCUDERO, A. GORDALIZA, AND C. MATRÁN (1999). A central limit theorem for multivariate generalized trimmed k-means. *Annals of Statistics*, **27**, 1061–1079.

L. A. GARCÍA-ESCUDERO, A. GORDALIZA, C. MATRÀN, AND A. MAYO-ISCAR (2008). A general trimming approach to robust cluster analysis. *Annals of Statistics*, **36**, 1324–1345.

L. A. GARCÍA-ESCUDERO, A. GORDALIZA, C. MATRÁN, AND A. MAYO-ISCAR (2010). A review of robust clustering methods. *Advances in Data Analysis and Classification*, **4**, 89–109.

L. A. GARCÍA-ESCUDERO, A. GORDALIZA, C. MATRÀN, AND A. MAYO-ISCAR (2011). Exploring the number of groups in robust model-based clustering. *Statistics and Computing*, **21**, 585–599.

L. A. GARCÍA-ESCUDERO, A. GORDALIZA, C. MATRÁN, AND A. MAYO-ISCAR (2014). Avoiding spurious local maximizers in mixture modeling. *Statistics and Computing*.

L. A. GARCÍA-ESCUDERO, A. GORDALIZA, R. SAN MARTIN, S. VAN AELST, AND R. ZAMAR (2009). Robust linear clustering. *Journal of the Royal Statistical Society (Series B)*, **71**, 301–318.

U. GATHER, S. KUHNT, AND J. PAWLITSCHKO (2003). Concepts of outlyingness for various data structures. In: J. C. MISRA, ed., *Industrial Mathematics and Statistics*, 545–585. Narosa Publishing House, New Delhi.

M. G. GENTON AND A. LUCAS (2003). Comprehensive definitions of breakdown points for independent and dependent observations. *Journal of the Royal Statistical Society (Series B)*, **65**, 81–94.

D. GERVINI (2002). The influence function of the Stahel-Donoho estimator of multivariate location and scatter. *Statistics and Probability Letters*, **60**, 425–435.

R. GNANADESIKAN AND J. R. KETTENRING (1972). Robust estimates, residuals, and outlier detection with multiresponse data. *Biometrics*, **28**, 81–124.

A. GORDALIZA (1991). Best approximations to random variables based on trimming procedures. *Journal of approximation theory*, **64**, 162–180.

A. GORDON (1999). *Classification*. Chapman and Hall.

J. GUO, G. JAMES, E. LEVINA, G. MICHAILIDIS, AND J. ZHU (2010). Principal component analysis with sparse fused loadings. *Journal of Computational and Graphical Statistics*, **19**, 930–946.

F. R. HAMPEL (1971). A general qualitative definition of robustness. *Annals of Mathematical Statistics*, **42**, 1887–1896.

F. R. HAMPEL (1974). The influence curve and its role in robust estimation. *Journal of the American Statistical Association*, **69**, 383–393.

F. R. HAMPEL, P. J. ROUSSEEUW, E. RONCHETTI, AND W. A. STAHEL (1986). *Robust Statistics: the approach based on the influence function*. Wiley.

J. HARDIN AND D. M. ROCKE (2004). Outlier detection in the multiple cluster setting using the minimum covariance determinant estimator. *Computational Statistics & Data Analysis*, **44**, 625–638.

J. HARDIN AND D. M. ROCKE (2005). The distribution of robust distances. *Journal of Computational and Graphical Statistics*, **14**, 928–946.

H. H. HARMAN AND W. H. JONES (1966). Factor analysis by minimizing residuals (minres). *Psychometrika*, **31**, 351–368.

J. A. HARTIGAN (1972). Direct clustering of a data matrix. *Journal of the American Statistical Association*, **67**, 123–129.

J. A. HARTIGAN (1975). *Clustering algorithms*. John Wiley & Sons, Toronto.

J. A. HARTIGAN (1978). Asymptotic distributions for clustering criteria. *Annals of Statistics*, **6**, 117–131.

J. A. HARTIGAN AND M. WONG (1979). Algorithm AS136: a k-means clustering algorithm. *Applied Statistics*, **28**, 100–108.

T. Hastie, R. Tibshirani, and J. Friedman (2009). *The elements of statistical learning.* Springer, New York.

R. Hathaway (1985). A constrained formulation of maximum likelihood estimation for normal mixture distributions. *Annals of Statistics*, **13**, 795–800.

D. M. Hawkins and G. J. McLachlan (1997). High-breakdown linear discriminant analysis. *Journal of the American Statistical Association*, **92**, 136–143.

X. He and W.K. Fung (2000). High breakdown estimation for multiple populations with applications to discriminant analysis. *Journal of Multivariate Analysis*, **72**, 151–162.

C. Hennig (2004). Breakdown point for maximum likelihood estimators of location-scale mixtures. *Annals of Statistics*, **32**, 1313–1340.

C. Hennig and T.F. Liao (2013). How to find an appropriate clustering for mixed-type variables with application to socio-economic stratification. *Journal of the Royal Statistical Society (Series C)*, **62**, 309–369.

S. Heritier, E. Cantoni, S. Copt, and M.-P. Victoria-Feser (2009). *Robust methods in biostatistics.* Wiley, Chichester, U.K.

J. L. Jr. Hodges (1967). Efficiency in normal samples and tolerance of extreme values for some estimates of location. In: *Proc. Fifth Berkeley Symp. Math. Statist. Probab.*, vol. 1, 163–186. Univ. California Press.

K. Hodges and J. Wotring (2000). Client typology based on functioning across domains using the CAFAS: implications for service planning. *Journal of Behavioral Health Services and Research*, **27**, 257–270.

J. L. Horn (1965). A rationale and test for the number of factors in factor analysis. *Psychometrika*, **30**, 179–185.

O. Hössjer and C. Croux (1995). Generalizing univariate signed rank statistics for testing and estimating a multivariate location parameter. *Journal of Nonparametric Statistics*, **4**, 293–308.

H. Hotelling (1936). Relations between two sets of variates. *Biometrika*, **28**, 321–377.

C. Hou, F. Nie, Y. Jiao, C. Zhang, and Y. Wu (2013). Learning a subspace for clustering via pattern shrinking. *Information Processing & Management*, **49**, 871–883.

R. H. Hoyle and J. L. Duvall (2004). Determining the number of factors in exploratory and confirmatory factor analysis. *Handbook of quantitative methodology for the social sciences*, 301–315.

Z. HUANG (1998). Extensions to the k-means algorithm for clustering large data sets with categorical values. *Data Mining and Knowledge Discovery*, **2**, 283–304.

P. J. HUBER (1964). Robust estimation of a location parameter. *Annals of Mathematical Statistics*, **35**, 73–101.

P. J. HUBER (1981). *Robust Statistics*. Wiley, New York.

P. J. HUBER (1985). Projection pursuit. *Annals of Statistics*, 435–475.

P. J. HUBER AND E.M. RONCHETTI (2009). *Robust Statistics, 2nd edition*. Wiley.

L. HUBERT AND P. ARABIE (1985). Comparing partitions. *Journal of Classification*, **2**, 193–218.

M. HUBERT AND S. ENGELEN (2004). Robust PCA and classification in biosciences. *Bioinformatics*, **20**, 1728–1736.

M. HUBERT, P. J. ROUSSEEUW, AND S. VAN AELST (2008). High-breakdown robust multivariate methods. *Statistical Science*, **23**, 92–119.

M. HUBERT, P. J. ROUSSEEUW, AND K. VAN DEN BRANDEN (2005). ROBPCA: a new approach to robust principal components analysis. *Technometrics*, **47**, 64–79.

M. HUBERT, P. J. ROUSSEEUW, AND S. VERBOVEN (2002). A fast method for robust principal components with applications to chemometrics. *Chemometrics and Intelligent Laboratory Systems*, **60**, 101–111.

M. HUBERT AND K. VAN DRIESSEN (2004). Fast and robust discriminant analysis. *Computational Statistics & Data Analysis*, **45**, 301–320.

C. J. HUBERTY (1994). *Applied discriminant analysis*. Wiley.

P. HUMBURG, D. BULGER, AND G. STONE (2008). Parameter estimation for robust HMM analysis of ChIP-chip data. *BMC Bioinformatics*.

D. JIANG, C. TANG, AND A. ZHANG (2004). Cluster analysis for gene expression data: a survey. *IEEE Transactions on Knowledge and Data Engineering*, **16**, 1370–1386.

R. A. JOHNSON AND D. W. WICHERN (2002). *Applied multivariate statistical analysis*. Prentice and Hall.

I. JOLLIFFE (2005). *Principal component analysis*. Wiley Online Library.

I. T. JOLLIFFE, N. T. TRENDAFILOV, AND M. UDDIN (2003). A modified principal component technique based on the LASSO. *Journal of Computational and Graphical Statistics*, **12**, 531–547.

H. F. KAISER (1958). The varimax criterion for analytic rotation in factor analysis. *Psychometrika*, **23**, 187–200.

L. KAUFMAN AND P. J. ROUSSEEUW (1987). Clustering by means of medoids. In: Y. DODGE, ed., *Statistical Data Analysis based on the L_1 Norm*, 405–416. Elsevier, Amsterdam.

L. KAUFMAN AND P. J. ROUSSEEUW (1990). *Finding groups in data*. Wiley.

S. KHAN AND A. AHMAD (2004). Cluster center initialization algorithm for k-means clustering. *Pattern Recognition Letters*, **25**, 1293–1302.

T. KHATUN (2009). Measuring environmental degradation by using principal component analysis. *Environment, Development and Sustainability*, **11**, 439–457.

D. KIBLER, D. W. AHA, AND M. K. ALBERT (1989). Instance-based prediction of real-valued attributes. *Computational Intelligence*, **5**, 51–57.

P. A. LACHENBRUCH (1975). *Discriminant analysis*. Hafner Press, New York.

P. A. LACHENBRUCH, C. SNEERINGER, AND L. T. REVO (1973). Robustness of the linear and quadratic discriminant function to certain types of nonnormality. *Communications in Statistics*, **1**, 39–56.

M. H. C. LAW, M. A. T. FIGUEIREDO, AND A. K. JAIN (2004). Simultaneous feature selection and clustering using mixture models. *IEEE transaction on pattern analysis and machine intelligence*, **26**.

S. X. LEE AND G. J. MCLACHLAN (2013). Model-based clustering and classification with non-normal mixture distributions. *Statistical Methods & Applications*, **22**, 427–454.

E. L. LEHMANN AND J. P. ROMANO (2004). Generalizations of the familywise error rate. *Annals of Statistics*, **33**, 1138–1154.

C. LENG AND H. WANG (2009). On general adaptive sparse principal component analysis. *Journal of Computational and Graphical Statistics*, **18**, 201–215.

G. LI AND Z. CHEN (1985). Projection-pursuit approach to robust dispersion matrices and principal components: primary theory and Monte Carlo. *Journal of the American Statistical Association*, **80**, 759–766.

F. LIANG, S. MUKHERJEE, AND M. WEST (2007). The use of unlabeled data in predictive modeling. *Statistical Science*, **22**, 189–205.

B. G. LINDSAY (1994). Efficiency versus robustness: the case for minimum Hellinger distance and related methods. *Annals of Statistics*, **22**, 1018–1114.

N. Locantore, J. S. Marron, D. G. Simpson, N. Tripoli, J. T. Zhang, K. L. Cohen, G. Boente, R. Fraiman, B. Brumback, and C. Croux (1999). Robust principal component analysis for functional data. *TEST*, **8**, 1–73.

H. Lopuhaä (1989). On the relation between S-estimators and M-estimators of multivariate location and covariance. *Annals of Statistics*, **17**, 1662–1683.

H. Lopuhaä and P. J. Rousseeuw (1991). Breakdown point of affine equivariant estimators of multivariate location and covariance matrices. *Annals of Statistics*, **19**, 229–248.

J. Lu, K. N. Plataniotis, and A. N. Venetsanopoulos (2003). Regularized discriminant analysis for the small sample size problem in face recognition. *Pattern Recognition Letters*, **24**, 3079–3087.

J. Macqueen (1967). Some methods for classification and analysis of multivariate observations. In: *Proceedings of the 5th Berkeley symposium on mathematical statistics and probability*, vol. 1, 281–297. University of California Press, Berkeley, CA.

S. C. Madeira and A. L. Oliveira (2004). Biclustering algorithms for biological data analysis: a survey. *IEEE/ACM transactions on computational biology and bioinformatics*, **1**, 24–45.

B. F. J. Manly (2005). *Multivariate statistical methods: a primer*. CRC Press.

M. Markatou (2000). Mixture models, robustness, and the weighted likelihood methodology. *Biometrics*, **56**, 483–486.

M. Markatou, A. Basu, and B. G. Lindsay (1998). Weighted likelihood equations with bootstrap root search. *Journal of the American Statistical Association*, **93**, 740–750.

R. Maronna and P. Jacovkis (1974). Multivariate clustering procedures with variable metrics. *Biometrics*, **30**, 499–505.

R. A. Maronna (1976). Robust M-estimators of multivariate location and scatter. *Annals of statistics*, **4**, 51–67.

R. A. Maronna (2005). Principal components and orthogonal regression based on robust scales. *Technometrics*, **47**, 264–273.

R. A. Maronna, R. D. Martin, and V. J. Yohai (2006). *Robust statistics: theory and methods*. Wiley, New York.

R. A. Maronna and V. J. Yohai (1995). The behavior of the Stahel-Donoho robust multivariate estimator. *Journal of the American Statistical Association*, **90**, 330–341.

R. A. MARONNA AND V. J. YOHAI (2008). Robust low-rank approximation of data matrices with elementwise contamination. *Technometrics*, **50**, 295–304.

R. A. MARONNA AND R. H. ZAMAR (2002). Robust estimates of location and dispersion for high-dimensional datasets. *Technometrics*, **44**, 307–317.

R. D. MARTIN AND R. H. ZAMAR (1993). Bias robust estimation of scale. *The Annals of Statistics*, **21**, 991–1017.

G. MATEOS AND G. B. GIANNAKIS (2010). Sparsity control for robust principal component analysis. In: *Signals, Systems and Computers (ASILO-MAR), 2010 Conference Record of the Forty Fourth Asilomar Conference on*, 1925–1929. IEEE.

S. W. MCKECHNIE, P. R. EHRLICH, AND R. R. WHITE (1975). Population genetics of Euphydryas butterflies. I. genetic variation and the neutrality hypothesis. *Genetics*, **81**, 571–594.

G. J. MCLACHLAN (1992). *Discriminant analysis and statistical pattern recognition*. Wiley, New York.

G. J. MCLACHLAN AND K. BASFORD (1988). *Mixture models: inference and applications to clustering*. Marcel Dekker, Inc., New York.

G. J. MCLACHLAN, R. W. BEAN, AND D. PEEL (2002). A mixture model based approach to the clustering of microarray expression data. *Bioinformatics*, **18**, 413–422.

G. J. MCLACHLAN AND D. PEEL (2000). *Finite mixture models*. Wiley, New York.

P. D. MCNICHOLAS AND T. B. MURPHY (2008). Parsimonious Gaussian mixture models. *Statistics and Computing*, **18**, 285–296.

G. M. MEROLA (2014). Least squares sparse principal component analysis: a backward elimination approach to attain large loadings. *Australian & New Zealand Journal of Statistics*, in press.

D. J. MILLER AND J. BROWNING (2003). A mixture model and EM-based algorithm for class discovery, robust classification, and outlier rejection in mixed labeled/unlabeled data sets. *IEEE Transactions on Pattern Analysis and Machine Intelligence*, **25**, 1468–1483.

G. W. MILLIGAN AND M. C. COOPER (1985). An examination of procedures for determining the number of clusters in a data set. *Psychometrika*, **50**, 159–179.

B. MIRKIN (2005). *Clustering for data mining: a data recovery approach*. Chapman & Hall/CRC.

N. NEYKOV, P. FILSMOSER, R. DIMOVA, AND P. NEYTCHEV (2007). Robust fitting of mixtures using the trimmed likelihood estimator. *Computational Statistics & Data Analysis*, **52**, 299–308.

M. R. OLIVEIRA, J. A. BRANCO, C. CROUX, AND P. FILZMOSER (2004). *Robust redundancy analysis by alternating regression*. Springer.

T.J. O'NEILL (1978). Normal discrimination with unclassified observations. *Journal of the American Statistical Association*, **73**, 821–826.

C. PARK, A. BASU, AND B. G. LINDSAY (2002). The residual adjustment function and weighted likelihood: a graphical interpretation of robustness of minimum disparity estimators. *Computational Statistics & Data Analysis*, **39**, 21–33.

A. M. PIRES AND J. A. BRANCO (2010). Projection-pursuit approach to robust linear discriminant analysis. *Journal of Multivariate Analysis*, **101**, 2464–2485.

G. PISON, P. J. ROUSSEEUW, P. FILZMOSER, AND C. CROUX (2003). Robust factor analysis. *Journal of Multivariate Analysis*, **84**, 145–172.

G. PISON AND S. VAN AELST (2004). Diagnostic plots for robust multivariate methods. *Journal of Computational and Graphical Statistics*, **13**, 310–329.

G. PISON, S. VAN AELST, AND G. WILLEMS (2002). Small sample corrections for LTS and MCD. *Metrika*, **55**, 111–123.

D. POLLARD (1981). Strong consistency of k-means clustering. *Annals of Statistics*, **9**, 135–140.

D. POLLARD (1982). A central limit theorem for k-means clustering. *Annals of Probability*, **10**, 919–926.

E. S. PRASSAS, R. P. ROESS, AND W. R. McSHANE (2007). Cluster analysis as a tool in traffic engineering. *Transportation Research Record: Journal of the Transportation Research Board*, **1566**, 39–48.

X. QI, R. LUO, AND H. ZHAO (2013). Sparse principal component analysis by choice of norm. *Journal of Multivariate Analysis*, **114**, 127–160.

R DEVELOPMENT CORE TEAM (2012). *R: A Language and Environment for Statistical Computing*. R Foundation for Statistical Computing, Vienna, Austria. ISBN 3-900051-07-0.

S. RAY AND B. G. LINDSAY (2005). The topography of multivariate normal mixtures. *Annals of Statistics*, **33**, 2042–2065.

D. REICH, A.L. PRICE, AND N. PATTERSON (2008). Principal component analysis of genetic data. *Nature Genetics*, **40**, 491–492.

C. Reilly, C. Wang, and M. Rutherford (2005). A rapid method for the comparison of cluster analyses. *Statistica Sinica*, **15**, 19–33.

A. C. Rencher and W. F. Christensen (2012). *Methods of multivariate analysis*. John Wiley & Sons.

M. Riani, A. C. Atkinson, and A. Cerioli (2009). Finding an unknown number of multivariate outliers. *Journal of the Royal Statistical Society (Series B)*, **71**, 447–466.

M. Riani, A. Cerioli, A. C. Atkinson, and D. Perrotta (2014a). Monitoring robust regression. *Electronic Journal of Statistics*, **8**, 646–677.

M. Riani, A. Cerioli, and F. Torti (2014b). On consistency factors and efficiency of robust S-estimators. *TEST*, **23**, 356–387.

M. Riani, D. Perrotta, and F. Torti (2012). FSDA: A MATLAB toolbox for robust analysis and interactive data exploration. *Chemometrics and Intelligent Laboratory Systems*, **116**, 17–32.

M. Romanazzi (1992). Influence in canonical correlation analysis. *Psychometrika*, **57**, 237–259.

P. J. Rousseeuw (1984). Least median of squares regression. *Journal of the American Statistical Association*, **79**, 851–857.

P. J. Rousseeuw (1985). Multivariate estimation with high breakdown point. *Mathematical statistics and applications*, **8**, 283–297.

P. J. Rousseeuw (1987). Silhouettes: a graphical aid to the interpretation and validation of cluster analysis. *Journal of Computational and Applied Mathematics*, **20**, 53–65.

P. J. Rousseeuw (1998). Maxbias curves. In: Banks D. Kotz S., Read C., ed., *Encyclopedia of Statistical Sciences*, vol. 3. Wiley, New York.

P. J. Rousseeuw and C. Croux (1992). Explicit scale estimators with high breakdown point. L_1-*Statistical Analysis and Related Methods*, **1**, 77–92.

P. J. Rousseeuw and C. Croux (1993). Alternatives to the median absolute deviation. *Journal of the American Statistical Association*, **88**, 1273–1283.

P. J. Rousseeuw and A. M. Leroy (1987). *Robust regression and outlier detection*. Wiley-Interscience, New York.

P. J. Rousseeuw and K. Van Driessen (1999). A fast algorithm for the minimum covariance determinant estimator. *Technometrics*, **41**, 212–223.

P. J. Rousseeuw and K. Van Driessen (2006). Computing LTS regression for large data sets. *Data mining and Knowledge Discovery*, **12**, 29–45.

P. J. ROUSSEEUW AND B.C. VAN ZOMEREN (1990). Unmasking multivariate outliers and leverage points. *Journal of the American Statistical Association*, **85**, 633–639.

P. J. ROUSSEEUW AND V. J. YOHAI (1984). Robust regression by means of S-estimators. In: J. FRANKE, W. HÄRDLE, AND D. MARTIN, eds., *Robust and nonlinear time series analysis*, 256–272. Springer.

D. W. RUSSELL (2002). In search of underlying dimensions: the use (and abuse) of factor analysis. *Personality and Social Psychology Bulletin*, **28**, 1629–1646.

C. RUWET, L. A. GARCÍA-ESCUDERO, A. GORDALIZA, AND A. MAYO-ISCAR (2012). The influence function of the TCLUST robust clustering procedure. *Advances in Data Analysis and Classification*, **6**, 107–130.

C. RUWET, L. A. GARCÍA-ESCUDERO, A. GORDALIZA, AND A. MAYO-ISCAR (2013). On the breakdown behavior of robust constrained clustering procedures. *TEST*, **22**, 466–487.

M. SALIBIAN-BARRERA, S. VAN AELST, AND G. WILLEMS (2006). Principal components analysis based on multivariate MM estimators with fast and robust bootstrap. *Journal of the American Statistical Association*, **101**, 1198–1211.

M. SALIBIAN-BARRERA AND V. J. YOHAI (2006). A fast algorithm for S-regression estimates. *Journal of Computational and Graphical Statistics*, **15**, 414–427.

J. SAUNDERS (1980). Cluster analysis for market segmentation. *European Journal of Marketing*, **14**, 422–435.

G. SCHWARZ (1978). Estimating the dimension of a model. *Annals of Statistics*, **6**, 461–464.

A. J. SCOTT AND M. J. SYMONS (1971). Clustering methods based on likelihood ratio criteria. *Biometrics*, **27**, 387–397.

S. SELIM AND M. ISMAIL (1984). k-means-type algorithms: a generalized convergence theorem and characterization of local optimality. *IEEE Transactions on Pattern Analysis and Machine Intelligence*, **6**, 81–87.

H. SHEN AND J. Z. HUANG (2008). Sparse principal component analysis via regularized low rank matrix approximation. *Journal of Multivariate Analysis*, **99**, 1015–1034.

G. SHEVLYAKOV AND P. SMIRNOV (2011). Robust estimation of the correlation coefficient: An attempt of survey. *Austrian Journal of Statistics*, **40**, 147–156.

W. A. STAHEL (1981). *Robuste schätzungen: infinitesimale optimalität und schätzungen von kovarianzmatrizen.* Ph.D. thesis, ETH Zürich.

D. STEWART AND W. LOVE (1968). A general canonical correlation index. *Psychological Bulletin*, **70**, 160.

M. SYMONS (1981). Clustering criteria and multivariate normal mixtures. *Biometrics*, **37**, 35–43.

G. M. TALLIS (1963). Elliptical and radial truncation in normal samples. *Annals of Mathematical Statistics*, **34**, 940–944.

A. TARSITANO (2003). A computational study of several relocation methods for k-means algorithms. *Pattern Recognition*, **36**, 2955–2966.

S. TASKINEN, C. CROUX, A. KANKAINEN, E. OLLILA, AND H. OJA (2006). Influence functions and efficiencies of the canonical correlation and vector estimates based on scatter and shape matrices. *Journal of Multivariate Analysis*, **97**, 359–384.

K. S. TATSUOKA AND D. E. TYLER (2000). On the uniqueness of S-functionals and M-functionals under nonelliptical distributions. *Annals of Statistics*, **28**, 1219–1243.

M. TENENHAUS (1998). *La régression PLS: théorie et pratique.* Editions Technip.

B. THOMPSON (2004). *Exploratory and confirmatory factor analysis: understanding concepts and applications.* American Psychological Association.

R. TIBSHIRANI, G. WALTHER, AND T. HASTIE (2001). Estimating the number of data clusters via the gap statistic. *Journal of the Royal Statistical Society (Series B)*, **63**, 411–423.

M. E. TIPPING AND C. M. BISHOP (1999). Mixtures of probabilitistic principal component analyzers. *Neural Computation*, **11**, 443–482.

V. TODOROV AND P. FILZMOSER (2009). An object oriented framework for robust multivariate analysis. *Journal of Statistical Software*, **32**, 1–47.

V. TODOROV AND A. M. PIRES (2007). Comparative performance of several robust linear discriminant analysis methods. *REVSTAT - Statistical Journal*, **5**, 63–83.

G. C. TSENG AND W. H. WONG (2005). Tight clustering: a resampling based approach for identifying stable and tight patterns in data. *Biometrics*, **61**, 10–16.

J. W. TUKEY (1962). The future of data analysis. *Annals of Mathematical Statistics*, **33**, 1–67.

S. VAN AELST AND P. ROUSSEEUW (2009). Minimum volume ellipsoid. *Wiley Interdisciplinary Reviews: Computational Statistics*, **1**, 71–82.

S. VAN AELST, E. VANDERVIEREN, AND G. WILLEMS (2011). Stahel-Donoho estimators with cellwise weights. *Journal of Statistical Computation and Simulation*, **81**, 1–27.

S. VAN AELST, E. VANDERVIEREN, AND G. WILLEMS (2012). A Stahel–Donoho estimator based on huberized outlyingness. *Computational Statistics & Data Analysis*, **56**, 531–542.

M. J. VAN DER LAAN, S. DUDOIT, AND K. S POLLARD (2004). Augmentation procedures for control of the generalized family-wise error rate and tail probabilities for the proportion of false positives. *Statistical Applications in Genetics and Molecular Biology*, **1**.

I. VAN MECHELEN, H-H. BOCK, AND P. DE BOECK (2004). Two-mode clustering methods: a structured overview. *Statistical Methods in medical research*, **13**, 363–394.

K. VARMUZA AND P. FILZMOSER (2008). *Introduction to multivariate statistical analysis in chemometrics*. CRC press.

H.D. VINOD (1969). Integer programming and the theory of grouping. *Journal of the American Statistical Association*, **64**, 506–519.

D. M. WITTEN, R. TIBSHIRANI, AND T. HASTIE (2009). A penalized matrix decomposition, with applications to sparse principal components and canonical correlation analysis. *Biostatistics*, **10**, 515–534.

M. WOLFSON, Z. MADJD-SADJADI, AND P. JAMES (2004). Identifying national types: a cluster analysis of politics, economics, and conflict. *Journal of Peace Research*, **41**, 607–623.

K. YEUNG, C. FRALEY, A. MURUA, A. RAFTERY, AND W. RUZZO (2001). Model-based clustering and data transformations for gene expression data. *Bioinformatics*, **17**, 977–987.

V. J. YOHAI AND R. A. MARONNA (1990). The maximum bias of robust covariances. *Communications in Statistics-Theory and Methods*, **19**, 3925–3933.

T. ZEWOTIR AND J. S. GALPIN (2007). A unified approach on residuals, leverages and outliers in the linear mixed model. *TEST*, **16**, 58–75.

B. ZHANG, M. HSU, AND U. DAYAL (2001). k-harmonic means: a spatial clustering algorithm with boosting. In: J. F. RODDICK AND K. HORNSBY, eds., *Temporal, Spatial, and Spatio-Temporal Data Mining*, 31–45. Springer.

H. ZOU AND T. HASTIE (2005). Regularization and variable selection via the elastic net. *Journal of the Royal Statistical Society (Series B)*, **67**, 301–320.

H. ZOU, T. HASTIE, AND R. TIBSHIRANI (2006). Sparse principal component analysis. *Journal of Computational and Graphical Statistics*, **15**, 265–286.

Y. ZUO, H. CUI, AND X. HE (2004). On the Stahel-Donoho estimator and depth-weighted means of multivariate data. *Annals of Statistics*, **32**, 167–188.

W. R. ZWICK AND W. F. VELICER (1986). Comparison of five rules for determining the number of components to retain. *Psychological Bulletin*, **99**, 432–442.

Index

affine equivariant, 43, 81, 192
applications, xi, 13, 59, 87, 109, 122, 138, 149, 172, 209, 210
approximation, 86
    χ^2, 55, 56, 67, 90
    scaled Beta, 52, 61
    scaled F, 51, 61
    Wilson-Hilferty, 86
ARE, 10, 31, 37

biplot, 73, 140, 144, 226
bisquare, 38, 47
Bonferroni, 56
breakdown point, 6, 10, 46, 153, 194
    asymptotic, 10
    bi-mode, 214
    cell, 11, 180
    finite sample, 10
    individual, 11, 176, 194, 216
    restricted, 11, 176, 194, 216
    universal, 11

C-step, 50, 174, 175, 192, 199
canonical correlation, 118
classification, xiii, 15, 24, 147, 219, 220
    semi-supervised, 147
cluster
    centroid, 14, 151, 164
    distribution, 159
    labels, 19
    orientation, 160
    shape, 14, 20, 147, 156, 160, 164, 172, 189
    volume, 160
    weights, 157
communality, 134

confidence interval, 1, 40
confusion matrix, 226
consistency, 32, 37, 175
    Fisher, 8
contamination, xi
    sensitivity, 12
cross validation, 222

distance, 150, 173
    Euclidean, 152, 172, 204, 211
    Mahalanobis, 45, 55, 156, 158, 220
    orthogonal, 86
    robust, 30, 46, 56
    score, 85
distance-distance plot, 61, 62
downweighting, xi

efficiency, 9
eigenvalues, 77, 118
eigenvectors, 77
elliptical, 29
empirical distribution function, 7
estimating equation, 32
    Huber, 32
    Tukey, 33
    unbiased, 32
exact fit, 84

factors, 133
false discovery exceedance, 57
false discovery rate, 57
forward search, 54
function
    Huber, 40
    influence, 6, 8, 30, 32, 37, 38, 43, 46, 153, 176, 193

267